Small-Molecule Modulators Targeting Emerging Therapeutic Pathways: Design, Synthesis and Biological Evaluation

Small-Molecule Modulators Targeting Emerging Therapeutic Pathways: Design, Synthesis and Biological Evaluation

Guest Editor

Yujun Zhao

MDPI

Basel • Beijing • Wuhan • Barcelona • Belgrade • Novi Sad • Cluj • Manchester

Guest Editor
Yujun Zhao
College of Pharmaceutical
Sciences
Zhejiang University of
Technology
Hangzhou
China

Editorial Office
MDPI AG
Grosspeteranlage 5
4052 Basel, Switzerland

This is a reprint of the Special Issue, published open access by the journal *Molecules* (ISSN 1420-3049), freely accessible at: https://www.mdpi.com/journal/molecules/special_issues/G0S39W11XZ.

For citation purposes, cite each article independently as indicated on the article page online and as indicated below:

Lastname, A.A.; Lastname, B.B. Article Title. *Journal Name* **Year**, *Volume Number*, Page Range.

ISBN 978-3-7258-6664-9 (Hbk)
ISBN 978-3-7258-6665-6 (PDF)
https://doi.org/10.3390/books978-3-7258-6665-6

Contents

About the Editor

Yujun Zhao

Yujun Zhao is a Professor of Medicinal Chemistry at the College of Pharmaceutical Sciences, Zhejiang University of Technology, Hangzhou, China. His research interests focus on medicinal chemistry and chemical biology, particularly the design and synthesis of small-molecule inhibitors and targeted protein degraders (PROTACs) against challenging protein–protein interactions implicated in cancer and inflammatory diseases, including MDM2/MDM4–p53, TEAD–YAP, RBM39-DCAF15, and NF-κB-inducing kinase. He has led multiple nationally funded research projects supported by the National Natural Science Foundation of China and strategic programs of the Chinese Academy of Sciences, advancing several translational drug discovery efforts. He has published extensively in leading journals such as *J. Am. Chem. Soc.*, *Cancer Cell*, and *J. Med. Chem.*, and is an inventor on numerous patents in small-molecule therapeutics. He currently serves as a Guest Editor or Editorial Board Member for international journals including *Molecules* and *Sci. Rep.*

Preface

This Reprint, entitled "Small Molecule Modulators Targeting Emerging Therapeutic Pathways Design Synthesis and Biological Evaluation", comprises eight peer-reviewed articles published in *Molecules* that collectively highlight recent advances in small molecule drug discovery. Despite major progress in biologics and gene-based therapies, small molecules remain indispensable therapeutic agents owing to their structural diversity, synthetic accessibility, favorable pharmacokinetic properties, and ability to modulate both enzymatic and non-enzymatic targets. The scope of this Reprint encompasses target identification and validation, rational molecular design, mechanistic elucidation, and biological evaluation, with a particular emphasis on emerging and non-canonical therapeutic pathways relevant to cancer, inflammation, and cardiovascular disease. Together, the contributions reflect contemporary strategies aiming to expand the druggable proteome and overcome resistance mechanisms associated with established therapies.

Two contributions are especially noteworthy for in-depth mechanistic insight and broad translational relevance. In their study of anticancer quinolinol small molecules, Wang and coauthors identified the molecular targets of MMRi62, a quinolinol-based compound frequently associated with PAINS-related concerns. By integrating structure–activity relationship analysis, chemical biology probe development, and quantitative proteomic profiling, the authors clarify the complex mode of action of MMRi62-like compounds. Importantly, this work demonstrates p53-independent, multi-pathway-mediated cell death in melanoma models, including those resistant to BRAF inhibitors, thereby providing mechanistic validation for a class of compounds often viewed with skepticism. These findings offer an important framework for the rational evaluation of polypharmacological small molecules and support their potential utility in addressing drug resistance in cancer therapy.

Sülzen and coauthors provide detailed structural insights into salinosporamide, a mediated inhibition of the human 20S proteasome. Using high-resolution cryo electron microscopy, the authors elucidate how this marine-derived γ-lactam β-lactone compound covalently modifies the catalytic subunits of the proteasome. The resulting structures define the molecular basis of irreversible proteasome inhibition and underscore the continued relevance of natural products as sources of structurally unique and mechanistically informative small molecule modulators.

The remaining contributions further broaden the scope of this Reprint by presenting integrated biophysical and computational analyses of tryptase inhibitors, studies on LXR-driven regulation of oxylipin formation during macrophage polarization, and the discovery of novel GLI1 inhibitors targeting Hedgehog signaling. Additionally, the articles review recent advances in nitro heteroaromatic compounds for hypoxia-selective cancer therapy and imaging, examine moonlighting enzymes as unconventional cancer targets, and summarize progress in naturally occurring PCSK9 inhibitors for cardiovascular disease. Collectively, these original research studies and comprehensive reviews reflect the multidisciplinary nature of modern small molecule discovery.

Overall, this Reprint highlights how advances in chemistry, structural biology, biophysics, and chemical biology are converging to enable the identification and modulation of emerging therapeutic pathways. It is intended for researchers in medicinal chemistry, chemical biology, and related biomedical fields who seek mechanistic insights and innovative strategies for small-molecule-based drug discovery.

Yujun Zhao
Guest Editor

molecules

MDPI

Article

Anticancer Quinolinol Small Molecules Target Multiple Pathways to Promote Cell Death and Eliminate Melanoma Cells Resistant to BRAF Inhibitors

Xinjiang Wang [1,*], Rati Lama [1], Alexis D. Kelleher [2], Erika C. Rizzo [2], Samuel L. Galster [2], Chao Xue [3], Yali Zhang [4], Jianmin Wang [4], Jun Qu [3] and Sherry R. Chemler [2,*]

1 Department of Pharmacology and Therapeutics, Roswell Park Comprehensive Cancer Center, Buffalo, NY 14263, USA
2 Department of Chemistry, University at Buffalo, State University of New York, Buffalo, NY 14260, USA
3 Department of Pharmaceutical Sciences, School of Pharmacy, University at Buffalo, State University of New York, Buffalo, NY 14260, USA
4 Department of Biostatistics and Bioinformatics, Roswell Park Comprehensive Cancer Center, Buffalo, NY 14263, USA
* Correspondence: xinjiang.wang@roswellpark.org (X.W.); schemler@buffalo.edu (S.R.C.)

Abstract: Small molecule inhibitors that target the E3 ligase activity of MDM2-MDM4 have been explored to inhibit the oncogenic activity of MDM2-MDM4 complex. MMRi62 is a small molecule that was identified using an MDM2-MDM4 E3 ligase-based high throughput screen and a cell-death-based secondary screen. Our previous studies showed that MMRi62 promotes MDM4 degradation in cells and induces p53-independent apoptosis in cancer cells. However, MMRi62 activity in solid tumor cells such as melanoma cells, especially in BRAF inhibitor resistant melanoma cells, have not been explored. Although its promotion of MDM4 degradation is clear, the direct MMRi62 targets in cells are unknown. In this report, we show that MMRi62 is a much more potent p53-independent apoptosis inducer than conventional MDM2 inhibitors in melanoma cells. A brief structure-activity study led to development of SC-62-1 with improved activity. SC-62-1 potently inhibits and eliminates clonogenic growth of melanoma cells that acquired resistance to BRAF inhibitors. We developed a pair of active and inactive SC-62-1 probes and profiled the cellular targets of SC-62-1 using a chemical biology approach coupled with IonStar/nano-LC/MS analysis. We found that SC-62-1 covalently binds to more than 15 hundred proteins in cells. Pathways analysis showed that SC-62-1 significantly altered several pathways including carbon metabolism, RNA metabolism, amino acid metabolism, translation and cellular response to stress. This study provides mechanistic insights into the mechanisms of action for MMRi62-like quinolinols. This study also suggests multi-targeting compounds like SC-62-1 might be useful for overcoming resistance to BRAF inhibitors for improved melanoma treatment.

Keywords: MMRi62; quinolinol; MDM2-MDM4; E3 ligase; cell death; targets; chemical robe

1. Introduction

Small-molecule inhibitors targeting cancer-driver protein kinases and enzymes constitute a majority of the targeted therapies approved by FDA in the past decades [1]. These targeted therapeutics are designed to target the interfaces in ATP binding pockets, domains of kinase activation, catalytic domains of enzymes or ligand binding for inactivation of the kinase activity that drives the cancerous cell proliferation. Although significantly improved cancer treatment for many cancer types, these therapies without exception face three challenges: low responsive rate, short-lived efficacy and drug resistance. Low responsive rate may be caused by additional genetic alterations in addition to the alterations in the intended drug targets in different patients. Short-lived efficacy is often intertwined with drug resistance that includes intrinsic and acquired drug resistance. The intrinsic drug resistance can be mediated by tumor cell heterogeneity, differential expression of efflux pump, and tumor microenvironment, while acquired resistance of initial responsive cancer cells can be gained through distinct mechanisms including gene mutation and non-genomic alteration of the targeted signaling pathways [2]. For example, acquired resistance to first and second-generation EGFR inhibitors in lung cancer can be conferred by point mutations such EGFR T790M mutation and resistance to third-generation EGFR inhibitors can be conferred by EGFR C797S mutation [3]. In the case of mutant BRAF inhibitors in melanoma, the acquired resistance is conferred by mechanisms involving genetic and epigenetic changes that activate different signaling pathways such as MAPK to bypass the effect of BRAF inhibition [4]. Likewise, inhibitors targeting MDM2-p53 interaction for p53-based therapies face resistance mechanisms involving mutations in p53 and other components of the p53 pathway [5–7] including MDM4 overexpression [8–10].

Despite our understanding of the various resistance mechanisms, the root cause of the limited efficacy of targeted therapies is lack of potent cell killing effect. This is because most drug targets for targeted therapies are mutated or deregulated kinases or enzymes, whose inactivation most likely results in reduced proliferation rather than cell death. While reduced proliferation prolongs progression-free survival, it allows cancer cells to rewire to alternative pathways or the treatment pressure selects resistant cancer populations. Eventually, the once-shrank tumors adapt the blockade by the targeted therapies leading to regrowth of more aggressive subtypes. In case of targeting MDM2 for p53-based cancer therapy, drug response to MDM2 inhibitors may differ significantly in terms of whether apoptosis or growth arrest is the major endpoint. This is because p53 plays a differential effect in apoptosis and cell cycle arrest in different cell types, with mainly apoptosis in leukemic or lymphoma cells and mainly growth arrest in fibroblasts and other cell types [11]. Indeed, MDM2 inhibitors induced a mixture of apoptosis and growth arrest in cancer patients in clinical trials [12,13].

To identify inhibitors that target the oncogenic activity of MDM2 with better cell killing activity, we used a unique approach for screening the lead compound. Our strategy was to first identify small molecules that target the RING domain of MDM2 which carries the E3 ligase activity and then use apoptosis assay in the secondary screen of primary hits identified by the biochemical screening [14]. This strategy is based on findings that the RING domains of both MDM2 and MDM4 are essential for p53 regulation *in vivo* [15–17] and that the E3 ligase activity of MDM2-MDM4 RING heterodimers is required for both p53 regulation *in vivo* and p53-independent cell cycle regulation [18]. Using this strategy, we identified MMRi62 that modulates the E3 ligase activity of MDM2-MDM4, promotes MDM4 degradation in cells and potently induces p53-independent apoptosis in leukemic cells [19] and ferroptosis in pancreatic cells [20]. Although MMRi62 hits the intended targets of MDM2-MDM4 complex, it is not clear how MMRi62 potently induces cancer cell death in different death mode. In this study, we report that MMRi62 has advantages over conventional MDM2 inhibitors in cell death induction in melanoma cells and SC-62-1 as MMRi62 derivatives has improved activity and is useful in eliminating melanoma cells with acquired resistance. We also report the results of SC-62-1-targeted pathways using SC-62-1 chemical probes and quantitative proteomics.

2. Results and Discussion

2.1. MMRi62 Has Advantage of Preferential Induction of Apoptosis over MDM2-p53 Inhibitors in Melanoma Cells

Our previous study showed that MMRi62 induces MDM4 degradation and p53-independent apoptosis in leukemic cells [19]. We asked if this pro-apoptotic activity of MMRi62 also takes place in solid tumor cells such as melanoma cells. We used A375 and p53-knockdown shp53-A375 cells and treated them with either MMRi62 or AMG232, a potent MDM2 inhibitor that specifically disrupts MDM2-p53 interaction [21] as a comparison in anti-proliferation assays. Our results showed that MMRi62 and AMG232 had comparable IC_{50} values in A375 cells, but MMRi62's anti-melanoma activity was not affected by p53 knockdown since A375 and shp53-A375 had similar IC_{50}s (Figure 1A,B). In contrast to MMRi62, AMG232 showed a 40-fold reduced activity in shp53-A375 cells as compared to A375 cells (Figure 1B). These data demonstrated that the anticancer activity of AMG232 is p53-dependent and MMRi62's activity is p53-independent, although both compounds induced p53 accumulation in A375 cells (Figure 1C,D). In contrast to AMG232, MMRi62-treated cells showed strong MDM4 downregulation in both A375 and shp53-A375 cells. MMRi62 treatment induced fast and strong apoptotic PARP cleavage (cPARP) in A375 cells at 24 h while AMG232 induced cPARP in A375 cells only after 48 h treatment (Figure 1C–E). MMRi62-induced cPARP was much weaker in shp53-A375 cells than in A375 cells, suggesting a major p53-dependent apoptosis mechanism was triggered in A375 cells. However, the MMRi62-induced p53-independent apoptosis was sustained up to 72 h in MMRi62-treated shp53-A375 cells (Figure 1C,F). Given that MMRi62 had similar IC_{50} values in A375 and shp53-A375 cells but with different levels of apoptosis, MMRi62's anti-melanoma activity must involve multiple death modes including p53-dependent apoptosis, p53-independent apoptosis and non-apoptotic cell death.

2.2. SC-62-1, an MMRi62 Derivative with Improved Activity

A previous SAR study on MMRi62 and MMRi67 identified the critical role of the quinolinol's hydroxyl moiety in anti-proliferation activity of these molecules [22]. Optimization of MMRi62 led to development of SC-62-1. We synthesized the methoxy counterparts of MMRi62 (MMRi62Me) and SC-62-1 (SC-62-1Me) and tested them together in A375 melanoma cells. Our results reached the same conclusion that hydroxyl group of MMRi62/SC-62-1 is essential for its activity since their methoxy counterparts lost activity (Figure 2). The calculated CLogP of SC-62-1 is 2.13, placing it in the realm of the mean value of ClogP for approved drugs, which has stayed consistently at 2.3–2.6 over a long period of time [23]. By replacing a chloro substituent with its bioisosteric nitrile [24] in the structure of SC-62-1, we have improved the water solubility over MMRi62 whose calculated CLogP is 3.42. To assess whether SC-62-1 is less toxic to non-cancer cells, we tested SC-62-1 in immortalized human keratinocytes HaCaT cells and obtained an IC_{50} of 0.61 ± 0.13 μM which is 5-fold higher than that of A375 cells, suggesting that SC-62-1 has limited degree of selectivity for melanoma cells over immortalized keratinocytes. We noticed that HaCaT cells grew as fast as A375 cells in culture which might suggest that SC-62-1 toxicity to HaCaT cells may be related to its cancer-like fast proliferation.

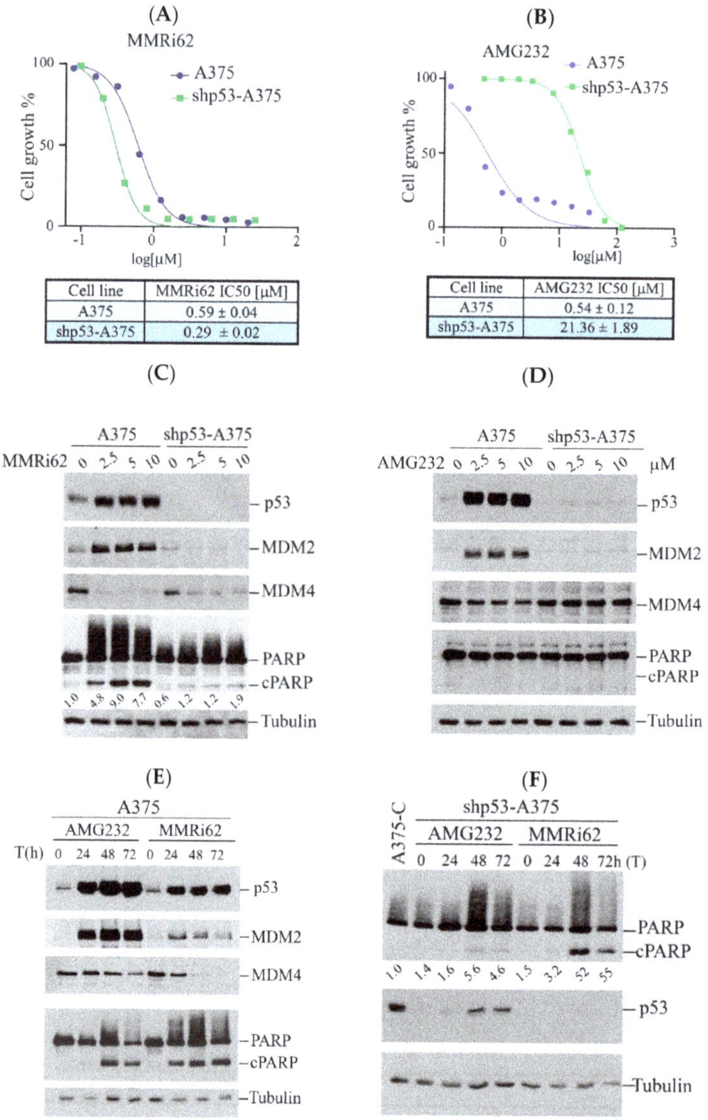

Figure 1. Westen Blot analysis of MMRi62 and AMG232 for the effect on MDM2, MDM4, p53 and cleaved PARP induction. (**A**,**B**) Dose response curves in the presence of MMRi62 (**A**) or AMG232 (**B**) for A375 and shp53-A375 cells. IC_{50}s are shown at the bottom. The representative data of three independent experiments were shown and the IC_{50} values were obtained from three independent experiments. (**C**,**D**) WB analysis of MDM2, MDM4, p53 and cleaved PARP (cPARP) induction in A375 and shp53-A375 cells treated with MMRi62 (**C**) or AMG232 (**D**) for 24 h (**E**). WB analysis of MDM2, MDM4, p53 and cleaved PARP (cPARP) induction in A375 treated with AMG232 or MMRi62 for 48 h and 72 h (**F**), WB analysis of cleaved PARP (cPARP) and p53 in shp53A375 cells treated with AMG232 or MMRi62 for 48 h and 72 h. Tubulin is the protein loading control. The intensities of cPARP and tubulin bands in (**C**,**F**) were quantified by ImageJ1.54P software and fold increases in cPARP over untreated control shown under cPARP WB were obtained after normalization against tubulin.

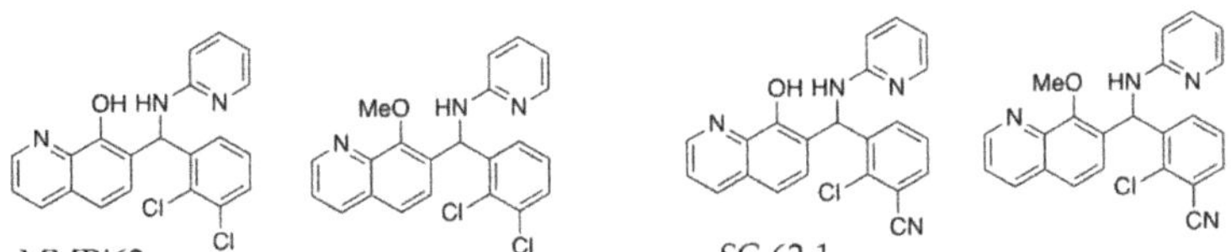

MMRi62
IC50 A375=
0.51 ± 0.03 μM

MMRi62Me
No activity

SC-62-1
IC50 A375=
0.11 ± 0.01 μM

SC-62-1Me
No activity

Figure 2. Chemical structure and antiproliferative activity of MMRi62, SC-62-1 and their inactive counterparts. Top: Chemical structures of MMRi62, MMRi62Me, SC-62-1 and SC-62-1Me. Bottom: The IC_{50} of MMRi62 and SC-62-1 in A375 cells.

2.3. SC-62-1 Prevents Appearance of Drug Resistance and Effectively Kills Melanoma Cells with Acquired Resistance to BRAF Inhibitors

Like most kinase inhibitors, BRAF inhibitors (BRAFi) such as Vemurafinib (Vem) potently inhibit growth of BRAFV600E mutant melanoma cells but do not eliminate them by cell death. A substantial population of resistant cells or drug-resistant persisters survives the BRAFi treatment and lead to disease progression [25,26]. To know how SC-62-1 affects the drug-resistant persisters in melanoma, we performed experiments with continuous drug exposure of A375 cells for 2 weeks with either Vem or SC-62-1 in cell culture by replacing the medium with fresh drug-containing medium every 3 days. The drug-resistant persisters always existed in Vem-treated plates. Vem up to 4 μM (33× IC_{50} dose) could not eliminate the persister cells. In contrast, SC-62-1 at 1 μM eliminated persisters (Figure 3A, microscopy images of crystal violet stained cells, top vs. middle). SC-62-1-Vem combination at 1:1 ratio eliminated Vem-resistant persisters at 2 and 4 μM but failed at 1 μM, suggesting synergism at doses higher than 2 μM. We readily established Vem-resistant cell lines (A375VemR cells) by a stepwise increase in Vem concentration in cell culture. The A375VemR cells acquired Vem resistance by a ~180-fold increase in the IC_{50} value. However, the A375VemR cells showed only little resistance to SC-62-1 with just a ~1.3-fold IC_{50} change in the IC_{50} value compared to A375 cells (Figure 3B). Importantly, A375VemR cells remained sensitive to SC-62-1. Exposure of A375VemR cells to SC-62-1 or SC-62-1-Vem combo at >4 μM for 8 days completely wiped out these cells, but Vem alone at 4-to-16 μM failed to do so (Figure 3C). The results of the antiproliferation assay indicated that the SC-62-1-Vem combinations generated strong synergisms in the A375VemR cells in a wide range of drug concentrations with inhibition levels higher than 40% (Figure 3D). The combination reached strong synergism level at effect levels >90% inhibition with a CI < 0.3. These data suggest that SC-62-1 can prevent the development of resistance and eliminate the resistant to Vem treatment cells on its own or with Vem at doses higher than 2 μM.

2.4. Design, Synthesis and Chracterization of SC-62-1 Probes

The capability of MMRi62 and SC-62-1 in inducing cancer cell death made them attractive tool compounds for uncovering the molecular basis of cancer cell killing. To identify MMRi62/SC-62-1's direct targets beyond the induced MDM4 degradation, we used an unbiased approach for profiling cellular targets with MMRi62/SC-62-1 probes coupled with quantitative proteomics.

Quinolinol Betti bases like MMRi62/SC-62-1 are synthesized via a three component Betti reaction involving condensation of 8-hydroxyquinoline, an aromatic aldehyde and 2-aminopyridine (Figure 4A). We hypothesized that quinolinols act as covalent inhibitors as their primary mechanism of action (MOA) via a retro-aza-Michael-quinone-methide mechanism [27,28]. The reactive quinone methides target cysteine side chains of cellular proteins for covalent bonding which may alter the enzymatic activity or structural function of their targets (Figure 4B). The potential targets of quinone methides can be broad including key enzymes in various pathways required for cancer cell growth and survival whose inactivation should trigger cell death due to cellular stress in different cancer types. This establishes a rationale that the quinolinol Betti bases may compensate the limitations of kinase targeted therapies by targeting multiple pathways to trigger cell death as an endpoint. In fact, about 30% of the marketed drugs are covalent inhibitors including tamoxifen and other targeted therapies [29–31]. To identify targets for covalent binding of SC-62-1 predicted by our hypothetical MOA, we designed and synthesized the SC-62-16/SC-62-16Me pair of probes where SC-62-16 is the active compound and SC-62-16Me is the inactive counterpart (Figure 4C). This pair of probes used an alkyl azide on C(5) for a "Click" reaction to crosslink the probe with alkyne-functionalized beads for pulldown of probe-bound proteins [32]. As expected, SC-62-16 was active in A375 cells with an IC_{50} of 1.36 μM and promotes MDM4/MDM2 degradation while SC-62-16Me was inactive with an IC_{50} of 134 μM without inducing MDM4/MDM2 degradation (Figure 4D,E). Therefore, SC-62-16Me is a good negative control for identifying off-targets of SC-62-1 not involved in anticancer mechanisms. As expected, downregulation of MDM4 by SC-62-16 is accompanied with p53 accumulation up to 5 μM. However, both MDM4 and p53 were downregulated by 10 μM SC-62-16. We speculate that lower concentrations of SC-62-16 below 5 μM induced specific effect on MDM4-p53 axis while high concentrations of SC-62-16 at 10 μM or above caused inhibition of both MDM4 and p53 expression. A moderate degree of apoptotic PARP cleavage was induced by 5 μM and 10 μM of SC-62-16 without clear dose-dependency, suggesting that apoptosis is only a part of involved death mechanisms and higher concentrations of the compound tend to trigger non-apoptotic endpoints.

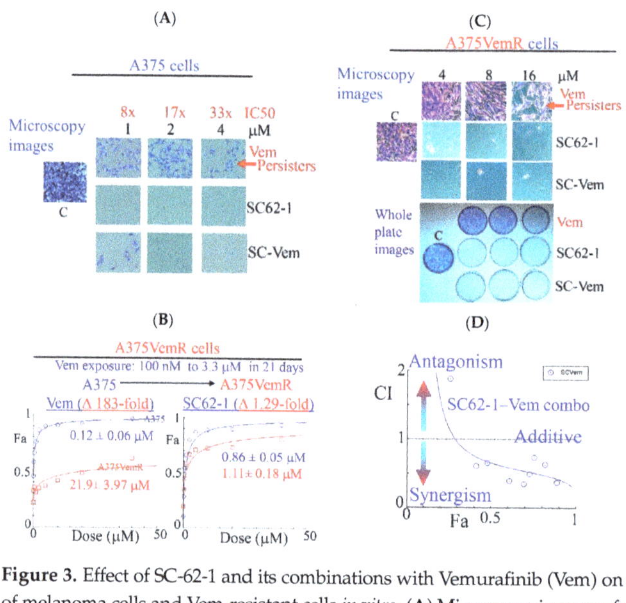

Figure 3. Effect of SC-62-1 and its combinations with Vemurafinib (Vem) on survival and proliferation of melanoma cells and Vem-resistant cells *in vitro*. (**A**) Microscopy images of crystal violet stained A375 cells after 2 week exposure to 1, 2, 4 μM Vem, SC-62-1 or their combinations. (**B**) Dose response curves of Vem-resistant A375 cells (A375VemR) with acquired resistance. (**C**) Microscopy images (top) and whole plates of crystal violet stained A375VemR cells after 8 day exposure to 4, 8, and 16 μM Vem, SC-62-1 or their combinations. (**D**) Combination index diagram obtained for SC-62-1-Vem combinations. CI < 1, synergism, CI = 1, additive, CI > 1, antagonism. Arrowheads indicate that the higher the CI values above 1, the more antagonistic they are, while the smaller the CI values below 1 toward zero, the more synergistic they are.

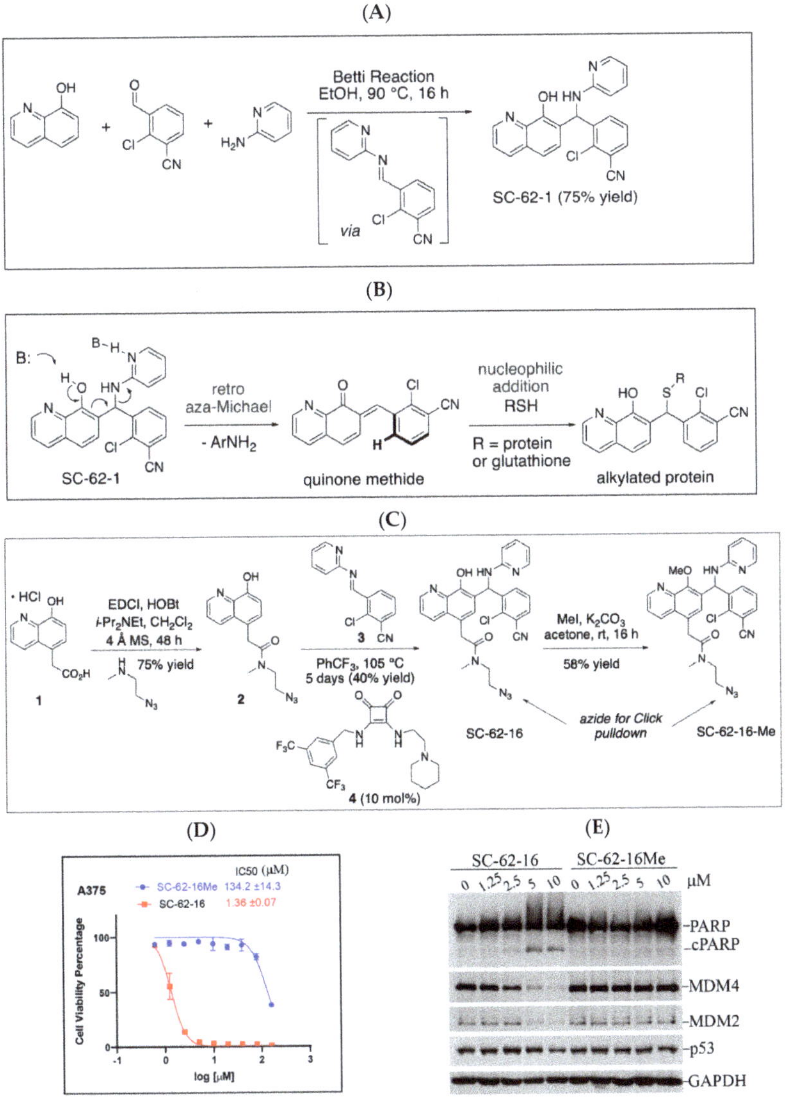

Figure 4. Synthesis of SC-62-1 and azide probes, hypothetical mechanisms of action of SC-62-1 and molecular characterization of azide probes. (**A**) Synthesis of quinolinol Betti base SC-62-1. (**B**) Hypothetical MOA for SC-62-1 as covalent inhibitors of target proteins via quinone methide to alkylate target proteins. (**C**) Synthesis routes for SC-62-16 and SC-62-16Me. (**D**) Anti-proliferation assays in A375 cells treated with SC-62-16 and SC-62-16Me for 72h. IC_{50s} values of the indicated compounds are shown. (**E**) Western blot analysis of SC-62-16 and SC-62-16Me effect on MDM4/MDM2 degradation and apoptotic PARP cleavage induction in treated A375 cells for 24 h. GAPDH is the protein loading control.

2.5. Identification of SC-62-16 Bound Proteins and Implicated Biological Pathways

We optimized a pulldown protocol in which biotin-PEG4-alkyne was used in click reaction [33,34] to tag a biotin moiety to SC-62-16/SC-62-16Me molecules that are covalently bound to proteins in treated cells, followed by pulldown of the biotin-tagged probe and its bound proteins by streptavidin-magnetic beads. The identities of proteins bound to the probes were then identified by quantitative IonStar/nano-LC/MS (Figure 5A). We treated A375 cells at 10 μM for 4 h before apoptosis occurs. Using 10 μM of the compounds ensured catching all possible SC-62-1 targets since this concentration induced strong MDM4 degradation and cell death at 24 h post-treatment (Figure 4E).

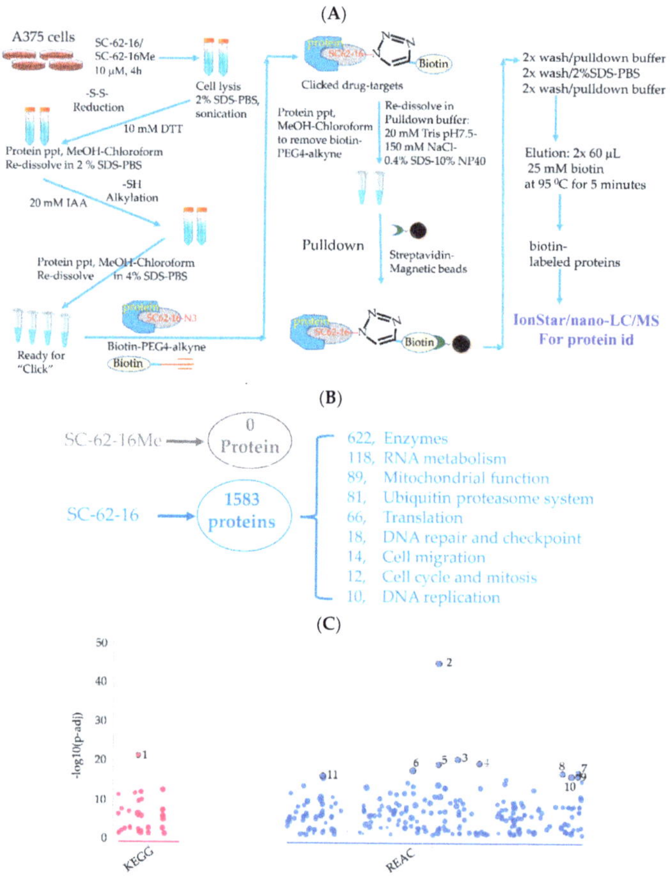

(D)

Id	source	term_id	term_name	term_size		p_value
1	KEGG	KEGG:01200	Carbon metabolism	115		5.5×10^{-22}
2	REAC	REAC:R-HAS-8953854	Metabolism of RNA	710		5×10^{-46}
3	REAC	REAC:R-HAS-86798695	Neutrophil degranulation	476		2.2×10^{-21}
4	REAC	REAC:R-HAS-72203	Processing of Capped Intron-Containing Pre-mRNA	276		2.4×10^{-20}
5	REAC	REAC:R-HAS-71291	Metabolism of amino acids and derivativs	370		4.4×10^{-20}
6	REAC	REAC:R-HAS-5663205	Infectious disease	1000		1.6×10^{-18}
7	REAC	REAC:R-HAS-72163	mRNA Splicing-Major pathway	202		4.3×10^{-18}
8	REAC	REAC:R-HAS-72766	Translation	292		8.1×10^{-18}
9	REAC	REAC:R-HAS-72172	mRNA Splicing	209		4.2×10^{-17}
10	REAC	REAC:R-HAS-9824446	Viral Infection Pathways	810		4.3×10^{-17}
11	REAC	REAC:R-HAS-2262752	Cellular responses to stress	792		5.1×10^{-17}

Figure 5. Identification of SC-62-16 bound proteins and targeted pathways. (**A**) Diagram for the optimized procedure of pulldown of SC-62-1-bound proteins. (**B**) Number of SC-62-16-bound proteins and their distributions in different categories. (**C**) Results of pathway analysis of the SC-62-16-bound proteins. The top 11 significantly altered pathways in using KEGG and REACTOME database are shown. (**D**) List of 11 pathways that are significantly impacted by SC-62-1.

Our approach worked well since the negative control compound SC-62-16Me did not bring down any proteins while SC-62-16 brought down more than 1500 proteins (Figure 5B). More than 600 of the SC-62-16-bound proteins are enzymes. These proteins are classified in several biochemical/biological categories including RNA metabolism, mitochondrial function, ubiquitin proteasome system, protein translation, DNA repair and checkpoint, cell migration, cell cycle and mitosis and DNA replication. When the sequences of these proteins were subjected to pathway analysis using KEGG [35] and REACTOME database [36], it was revealed that the significantly altered pathways include those for carbon metabolism, RNA metabolism, amino acid metabolism, RNA splicing, protein translation and cellular stress response (Figure 5C,D).

When the SC-62-16-bound proteins were grouped by biological processes or functions in a common pathway and ranked by the protein numbers of each group, they could be classified into six groups as shown in Table 1. RNA metabolism was in the number one ranked category with a total of 124 proteins including 87 RNA binding proteins such as SUB1, FUS, EWSR1, RBMX, RRP43, DDX17, . . .etc. (Table 1), and 37 proteins in mRNA splicing including SF3B1, SF3A1, RBM39, SF3B4, PRPF8, U2AF1, . . .etc. (Table 1).

The second highest group impacted mitochondrial function, with a total of 94 proteins including 50 metabolic enzymes, namely, NDUFV1, MTHFD2, PYCR1, MCAT, GPD2, GSR, PPA2, . . .etc. (Table 1), and 44 proteins for mitochondrial tRNA metabolism and mitochondrial maintenance including VARS2, VARSL, MARS2, APEX1, HAP1, IARS2, WARS2, AFG3L2, MRM1, CARS2, REXO2, . . .etc. (Table 1).

Table 1. SC-62-16 covalent binding proteins.

Pathway Ranking	Pathway Name	Proteins Number	PROTEIN Names (Some Are Partial List)
1	RNA metabolism	124	87 RNA binding proteins such as SUB1, FUS, EWSR1, RBMX, RRP43, DDX17,
			DDX9, ADAR, SRB, RNH1, LUC7L2, ILF3, UBAP2L, SERBP1, KHDRBS1, NONO,
			IGF2BP2, SRRT, FTSJ3, MED23, RBM4, RCL1, EXOSC3, SYNCRIP, DHX30, EBP2,
			REXO2, ZFR, DDX1, . . .
			37 mRNA splicing such as SF3B1, SF3A1, RBM39, SF3B4, PRPF8, U2AF1, SCAF11,
			SF1, SFRS19, SRSF9, SRSF4, SFRS18, SF4, SFRS14, SFPQ, SRSF1, SRSF2, SRSF7,
			SRSF3, SPF45, SAP145, PRPF4, SRRM2, . . .
2	Mitochondrial function	94	50 metabolic enzymes such as NDUFV1, MTHFD2, PYCR1, MCAT, GPD2, GSR,
			PPA2, PC, PCCA, MDH2, SHMT2, GOT2, ECHS1, IDH3A, CS, MCCC1, UQCRC1,
			SUCLG1, ALDH2, ECH1, HAGH, FH, IVD, ACAT1, ACADVL, DBT,
			SUCLG2, PDHB, CPOX, DLAT, AK2, PDHA1, COQ7, PDHX, HIBADH,
			SDHB, PRDX3, ATP5F1A, UQCRC2, ATP5F1B, NDUFS1, PGAM5,
			ATP5PB, HSD17B10, HADHB, GLS, ATP5PO, DLST, MECR, OGDH,. . ..
			10 for mitochondrial tRNA metabolism: VARS2, VARSL, MARS2, IARS2,
			MRM1, CARS2, REXO2, TARS2, RPUSD3, WARS2.
			34 proteins for mitochondrial maintenance: APEX1, HAP1, WARS2, AFG3L2,
			TOMM70, PMPCB, CLPB, HSPD1, HSPA9, SLC25A24, SCAMC1, MTCH2,
			TUFM, ETFA, LRPPRC, SLC25A3, SSBP1, SLC25A22, HSPE1, DAP3, VDAC2,
			VDAC1, MRPS27, CCDC51, MTX1, MRPS5, TMEM126B, TMEM11, IMMT,
			MINOS2, LETM1, MAVS, CYCS, DNM1L.

Table 1. *Cont.*

Pathway Ranking	Pathway Name	Proteins Number	PROTEIN Names (Some Are Partial List)
3	Protein degradation	55	18 ubiquitin E3 ligases:
			TRIM25, ARIH1, ARIH2, RNF20, CBL, HECTD1, HUWE1, TRIM18, RAD18,
			RBBP6, RNF113A, RNF149, RNF220, RNF34, RNFT2, TRIP12, UBR1, UBR4
			5 ubiquitin-conjugating enzymes: UBE2D2, UBE2D3, UBE2L3, UBE2N, UBE2Z,
			4 deubiquitinases: USP47, USP5, USP7, USP9
			28 proteasome components: PSMD4, PSMB2, PSMB6, PSMB4, PSMB1, ECPAS,
			PSMA6, PSMA1, PSMB7, PSMD2, PSMA3, PSMD7, PSMD13, PSMA2, PSMC4,
			PSMC6, PSMD8, PSMB5, PSMA5, PSMD11, PSMC3, PSMD1, PSMC1, PSMD4,
			PSMA7, PSMC2, PSMA4, PSMD12.
4	Protein translation	50	18 translation initiation factors: EIF3A, EIF6, EIF3G, EIF5A, EIF4G1, EIF5, EIF3I,
			EIF2S2, CDC123, EIF5B, EIF2S3, EIF3D, EIF3H, EIF3B, EIF3K, EIF4G3, EIF4G2,
			EIF4B.
			22 tRNA ligases, RARS1, VARS1, YARS1, AARS1, IARS1, EPRS1, TARS2,
			QARS1, HARS1, MARS1, WARS2, FARSB, IARS2, MARS2, AARS2, TARS1,
			SARS1, CARS1, NARS1, GARS1, WARS1, FARSA.
			10 ribosome proteins: GCN1, RRBP1, BMS1, NIP7, BOP1, MRTO4, LRRC59,
			GCN1, TMA7B, EEF1D.
5	DNA repair and	18	RAD50, RAD23B, TREX1, MSH6, ERCC6L, RAD18, ERCC5, XRCC6, MRE11,
	checkpoint response		APEX1, MMS19, BUB3, SMC2, MDC1, TP53BP1, FANCI, BRAT1, TOP1.
6	Cell migration	12	AGRIN, RAC1, MIG8, NUDC, MIG10, MYH9, RHOC, GIT1, ARHGAP34,
			ARHGAP17, ARHGDIA, MYL6B.

The third group targets protein degradation, with a total of 55 proteins that include 18 ubiquitin E3 ligases, TRIM25, ARIH1, ARIH2, RNF20, CBL, HECTD1, HUWE1, TRIM18, RAD18, RBBP6, RNF113A, RNF149, RNF220, RNF34, RNFT2, TRIP12, UBR1, UBR4, 5 E2 ubiquitin-conjugating enzymes, UBE2D2, UBE2D3, UBE2L3, UBE2N, UBE2Z, 4 deubiquitinases, USP47, USP5, USP7, USP9 and 28 proteasome components including PSMD4, PSMB2, PSMB6, PSMB4, PSMB1, ECPAS, PSMA6, PSMA1, PSMB7, PSMD2, PSMA3, PSMD7, PSMD13, PSMA2, PSMC4, PSMC6, PSMD8, PSMB5, . . .etc. (Table 1).

The fourth group impacts protein translation, with a total of 50 proteins including 18 translation initiation factors including EIF3A, EIF6, EIF3G, EIF5A, EIF4G1, EIF5, . . .etc. (Table 1), 22 tRNA ligases including RARS1, VARS1, YARS1, AARS1, IARS1, EPRS1, TARS2, QARS1, . . .etc. (Table 1), and 10 ribosome proteins including GCN1, RRBP1, BMS1, NIP7, BOP1, MRTO4, LRRC59, GCN1, TMA7B, EEF1D. The fifth group contains 18 proteins involved in DNA repair and the DNA damage checkpoint response (RAD50, RAD23B, TREX1, MSH6, ERCC6L, RAD18, ERCC5, XRCC6, MRE11, APEX1, MMS19, BUB3, SMC2, MDC1, TP53BP1, FANCI, BRAT1, TOP1) and the last group is involved in cell migration and contains 12 proteins (AGRIN, RAC1, MIG8, NUDC, MIG10, MYH9, RHOC, GIT1, ARHGAP34, ARHGAP17, ARHGDIA, MYL6B) (Table 1).

Although our findings suggest that SC-62-1 alkylates ~1500 proteins in cells, the effect of alkylation on individual proteins is difficult to assess without further study of their effect on downstream components of their involved pathways. We speculate that the effect on the function of individual proteins can vary significantly and not all these potential alkylation events have negative impact on the function of target proteins. For example, there were 28 proteasome components among the targets, the collective effect of their inactivation on the proteasome activity could be significant and result in severe defect in proteasomal degradation of ubiquitinated proteins such as MDM4. However, MMRi62 and SC-62-16 but not SC-16Me induced fast proteasomal degradation of MDM4 as we showed in this study (Figures 1 and 4E). These results suggested that the proteasome function was not defective at least for the induced proteasomal degradation of MDM4, despite that many of the proteasome components were alkylated by MMRi62 and SC-62-16. This suggests that while this target profile of SC-62-1 provides a comprehensive list of potential drug targets, only a limited number of protein targets have a meaningful consequence upon alkylation and are responsible for the drug effect. It appears that SC-62-1-mediated alkylation does not occur by random contact with SH-group since simply mixing SC-62-1 and glutathione in deuterated $DMSOd_6$ and heating at 37 °C for 24 h did not produce any change in the ^1H NMR spectrum of SC-62-1, which may indicate that alkylation occurs primarily inside proteins, possibly facilitated by acid/base catalysis and "intramolecularity".

A genetic screen for enriched genes in cells surviving SC-62-1 treatment might help sort out the targets that are responsible for the drug response [37]. Nevertheless, we speculate that SC-62-1 must have inactivated multiple key proteins and touched multiple pathways that prevent cells from overcoming the collateral damage. Based on the phenotypical effect of cell killing and the target profile of SC-62-1, we propose a working model as a possible mechanism of action for SC-62-1, shown in Figure 6. This model proposes that SC-62-1 acts as a multi-targeting alkylating agent. Through alkylating its target proteins, SC-62-1 not only induces proteasomal degradation of MDM4 and lysosomal degradation of FTH1, but also alters multiple pathways including RNA metabolism, mitochondrial function, protein translation and regulated protein degradation by proteasomes. The collective effect of its action creates overwhelming stressful state that cancer cells cannot overcome leading to p53-independent cell death in form of apoptosis in apoptosis-default cell types such leukemic cells or ferroptosis in ferroptosis-default cell types such as pancreatic and melanoma cells.

In summary, this study reports that quinolinol derivatives like MMRi62 and SC-62-1 preferentially induce cancer cell death compared to specific MDM2 inhibitors and BRAF inhibitors. They induce proteasomal degradation of MDM4 [19] and lysosomal degradation of FTH1 [20] but also target multiple other fundamental pathways for cell death induction. These compounds lack the specificity of conventional targeted therapies, a weakness measured by the concept of targeted therapies. However, pursuit of high specificity comes at the cost of short-lived efficacy and drug resistance, which is a well-established phenomenon for all targeted therapies. This limitation of targeted therapies lies in the capability of cancer cells easily bypassing the blockade of very specific drugs by mutating the well-defined interface or rewiring the affected signaling pathways. In fact, Mencher and Wang decades ago proposed that promiscuity of drugs can be a virtue compared to selective drugs [38]. Promiscuity has also been explored to overcome drug resistance to HIV-1 protease inhibitors [39]. In a recent study, 54 of 62 FDA-approved kinase inhibitors including Gleevec were reported to be promiscuous inhibitors with loose specificity of targeting at least two or more kinases. More interestingly, the study found that the more promiscuous type II inhibitors showed better pharmacodynamics than more specific type I inhibitors [40]. Therefore, it is not the specificity but the general toxicity and pharmacodynamics of a compound that determine its favorable therapeutic window. The toxicity of SC-62-1 is acceptable since its maximum tolerated dose in Scid mice is 60 mg/kg given every other day by intraperitoneal (I.P.) injection. We tested SC-62-1 for its *in vivo* efficacy in monotherapy in subcutaneous models of melanoma A375Luc in Scid mice at 57 mg/kg I.P. every other day and achieved ~50% reduction in tumor burden with body weight changes within 5%. However, due to small mouse number and large individual variation in tumor size, the results did not reach statistical significance. Future studies with a larger number of mice and development of derivatives with improved cancer cell selectivity will inform the potential of these quinolinol compounds as a new class of anticancer agents.

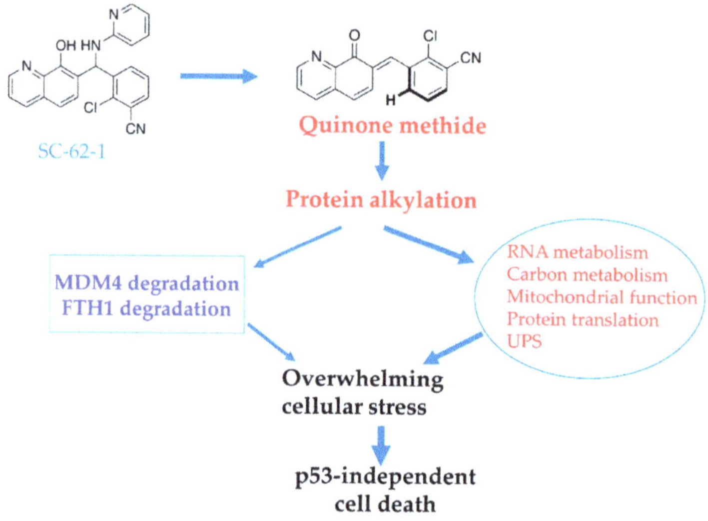

Figure 6. Proposed model for MOA of SC-62-1 in induction of cancer cell death. SC-62-1's anticancer activity is via production of quinone methide which alkylates many target proteins. Its downstream effect includes induction of proteasomal degradation of MDM4 protein and lysosomal degradation of FTH1 and significant alteration of the pathways including RNA metabolism, carbon metabolism, mitochondrial function, protein translation and protein degradation. Collectively, induction of changes in these pathways by SC-62-1 creates overwhelming cellular stress that cannot be resolved, thus leading to p53-independent cell death.

3. Materials and Methods

3.1. Representative Chemistry Methods

(±)-2-Chloro-3-((8-hydroxyquinolin-7-yl)(pyridin-2-ylamino)methyl) benzonitrile (SC-62-1)

To a 250 mL round bottomed flask equipped with a magnetic stir bar, 2-chloro-3-cyanobenzaldehyde (1.00 g, 6.04 mmol) was added along with 2-aminopyridine (568 mg, 6.04 mmol). 50 mL of absolute ethanol was added to the flask and the mixture was stirred until solids fully dissolved, at which point 8-hydroxyquinoline (1.05 g, 7.25 mmol) was added. The reaction flask was heated to 90 °C and refluxed for 24 h during which an off-white solid precipitated. The solution was allowed to cool to room temperature and to stand for several hours, then the resulting solid was isolated by filtration, giving **SC-62-1** (1.75 g, 75% yield) as an off-white powder. Mp = 190–192 °C; ^{1}H NMR (300 MHz, $CDCl_3$) δ 8.77 (d, J = 3.0 Hz, 1H), 8.13 (d, J = 7.2 Hz, 1H), 8.09 (d, J = 4.1 Hz, 1H), 7.97 (d, J = 7.9 Hz, 1H), 7.59 (d, J = 6.4 Hz, 1H), 7.50–7.26 (m, 6H), 6.70 (d, J = 6.4 Hz, 1H), 6.66–6.57 (m, 1H), 6.36 (d, J = 8.4 Hz, 1H), 5.55 (d, J = 6.2 Hz, 1H); ^{13}C NMR (75 MHz, $CDCl_3$) δ 157.2, 149.8, 148.3, 141.5, 138.2, 137.7, 136.1, 136.0, 133.0, 132.9, 127.9, 127.4, 127.0, 122.2, 120.6, 117.9,

116.2, 114.3, 114.0, 107.1, 53.6; IR (neat film): 3346, 2923, 2233, 1600, 1571, 1517, 1501 cm^{-1}; HRMS (ESI) calculated for $C_{22}H_{16}ClN_4O$ $[M + H]^+$: 387.1016, found 387.1007.

(±)-*N*-(2-azidoethyl)-2-(7-((2-chloro-3-cyanophenyl)(pyridin-2-ylamino)methyl)-8-hydroxyquinolin-5-yl)-*N*-methylacetamide (SC-62-16)

(Note: The 3-component Betti reaction above did not work for this 5-substituted-8-hydroxy quinoline, it is much less reactive than 8-hydroxy quinoline). To a 10 mL reaction tube, the imine **3** (42 mg, 0.18 mmol) was added, followed by the amide **2** (40 mg, 0.14 mmol, 0.8 eq.) dissolved in $PhCF_3$ (0.5 mL). The squaramide catalyst **3** (8 mg, 0.039 mmol, 0.1 eq.) was also added to the reaction tube. Flame-dried 4 Å molecular sieves (30 mg) were added, and the reaction tube was sealed and kept in a dark environment at 105 °C for 5 days. The reaction mixture was cooled to rt, diluted with CH_2Cl_2, and filtered through Celite then concentrated to a crude mixture. The crude was then recrystallized from MeOH to yield 42 mg of **SC-62-16** as light brown crystals in 40% yield. The NMR data indicates the presence of amide rotamers due to hindered rotation. Mp = 162–164 °C; ^1H NMR (400 MHz, $CDCl_3$) δ 8.79 (dd, *J* = 4.2, 1.5 Hz, 1H), 8.36 (ddd, *J* = 14.3, 8.6, 1.5 Hz, 1H), 8.08 (ddd, *J* = 5.0, 1.9, 0.8 Hz, 1H), 7.94 (dt, *J* = 7.9, 2.5 Hz, 1H), 7.59 (dd, *J* = 7.7, 1.6 Hz, 1H), 7.49 (ddd, *J* = 8.6, 4.2, 1.6 Hz, 1H), 7.45–7.30 (m, 2H), 7.25 (s, 1H), 6.69 (dd, *J* = 6.5, 2.4 Hz, 1H), 6.61 (ddd, *J* = 7.2, 5.0, 0.9 Hz, 1H), 6.38 (d, *J* = 8.4 Hz, 1H), 5.54 (d, *J* = 6.5 Hz, 1H), 4.08–3.91 (m, 2H), 3.51 (t, *J* = 5.4 Hz, 2H), 3.47–3.33 (m, 3H), 3.07 (s, 2H), 2.92 (s, 1H); ^{13}C NMR (101 MHz, $CDCl_3$) δ 170.5, 157.8, 149.3, 148.0, 147.8, 142.9, 138.5, 136.8, 135.4, 133.7, 133.4, 132.8, 127.9, 127.5, 122.5, 121.7, 120.8, 115.9, 113.8, 113.0, 108.8, 51.8, 51.7, 49.2, 48.7, 48.5, 47.0, 37.2, 36.7, 35.8, 32.5; HRMS (ESI) calculated for $C_{27}H_{24}ClN_8O_2$ $[M + H]^+$: 527.1711, found 527.1704.

The full synthesis and characterization details of all new compounds are provided in the Supplemental Information (SI).

3.2. Biological Assays and Methods

3.2.1. Cell Culture

Melanoma cell line A375 was cultured in DMEM-10% fetal bovine serum, 50 U/mL penicillin and 50 μg/mL streptomycin. Cell line with stable p53 knockdown shp53-A375 was established using pLKO.1-p53 (purchased from Addgene) (Plasmid #19119) [41] followed by puromycin selection at 1 μg/mL for 2 days, then clonal expansion in puromycin-free medium.

3.2.2. Western Blotting Analysis and Crystal Violet Staining

The Western blotting procedure and the relevant antibodies was described previously [19]. For crystal violet staining, cell plates were fixed with 4% (*w*/*v*) paraformaldehyde for 15 min followed by 30 min of staining in 0.1% (*w*/*v*) crystal violet solution in 20% (*v*/*v*) Ethanol-80% (*v*/*v*) PBS. After rinsing with tap water, the plates were airdried and the images were taken on Olympus IX73 inverted fluorescence microscope at a magnitude of 20×. The images of whole plates were taken on iPhone.

3.2.3. IC_{50} Measurement and Analysis

The procedure for measuring IC_{50} was described previously [19]. Some data were obtained by Chou-Median-Effect Equation using CompuSynversion 1.0 (CompuSyn.exe) software [42] and some were obtained by GraphPad (Prism 8) using affected fractions of compound-treated wells normalized against no-drug control wells with non-linear regression model.

3.2.4. Affinity Pulldown and Identification of SC-62-16/SC-62-16Me Covalently Bound Proteins

The procedure uses 2% (*w*/*v*) SDS-PBS lysis buffer to lyse the treated cells followed by treatment with 10 mM DTT, at 70 °C for 15 min to dissociate all non-covalent protein–protein interaction and reduce all the disulfide bonds among peptides. Then, the proteins were precipitated with 100% (*v*/*v*) methanol and 20% (*v*/*v*) chloroform, then were re-dissolved in 2% (*w*/*v*) SDS-PBS buffer by sonication. Before the "click" reaction, the samples were treated with iodoacetamide to alkylate sulfhydryl groups from further forming disulfide bonds. After "click" reaction with Biotin-PEG4-alkyne, the biotin-tagged compound-bound proteins were pulled down with magnetic Streptavidin beads. The bead-bound proteins were then eluted by 25 mM biotin heated at 95 °C for 5 min.

The click reaction conditions to link Biotin-PEG4-alkyne to SC-62-16 and SC62-16Me azide were adapted from a report by A. J. Pradhan et al. [43]. The conditions for biotin-streptavidin lysate preparation and biotin-streptavidin affinity purification and elution of compound-bound proteins were adapted from the method reported by J. S. Cheah et al. [44]. Biotin-PEG4-alkyne was purchased from Sigma-Aldrich (St. Louis, MO, USA) (cat# 764213-5MG) and PuroMAG™ Magnetic Beads-Streptavidin were purchased from Luna Nanotech (Markham, ON, Canada, Cat# MGB-STRP-10). Briefly, A375 cells (20 × 10 cm plates at 90% confluency) were treated with 10 μM SC-62-16 or SC-62-16Me for 4 h. After collecting the cells in a 50 mL centrifuge tube, they were washed with PBS and resuspended in 25 mL PBS and added 20% (*w*/*v*) SDS to final 2% (*w*/*v*) SDS-PBS followed by 20 cycles of sonication (30″-on-30″-off at 30% intensity of output) (Fisher Scientific Sonic Dismembrator Model 100 from Fisher Scientific, Waltham, MA, USA). Then, the cell lysate was treated with 10 mM DTT at 70 °C for 15 min. After centrifugation of the lysate at 50,000× *g* at 20 °C for 2 h, the soluble proteins were precipitated with 1× volume of methanol and 1/5× volume of chloroform with 5 min vigorous vortexing and 10 min centrifugation. After re-dissolving the proteins in 2% (*w*/*v*) SDS-PBS aided with sonication, the soluble protein solution was treated with 20 mM iodoacetamide for 30 min in darkness. Then, the proteins were precipitated again with methanol-chloroform and washed with cold methanol and resuspended in 4% (*w*/*v*) SDS-PBS. The 10× click reaction stock reagents were then added, to make a final concentration of 1 mM $CuSO_4$, 1 mM TCEP, 0.25 mM TBTA and 0.25 mM Biotin-PEG4-alkyne. After a brief vortex, the mixture was incubated at 37 °C for 1 h. Then, the proteins were precipitated with methanol-chloroform followed by two washes with methanol with sonication. The proteins were re-dissolved in 3 mL of biotin-streptavidin lysis buffer (50 mM Tris HCl pH 7.4, 150 mM NaCl, 0.4% (*w*/*v*) SDS, 1% (*v*/*v*) NP, 1 mM EGTA and 1.5 mM $MgCl_2$) and underwent 10 cycles of 30 secs-on-30 secs-off sonication in Sonic Dismembrator followed by centrifugation at maximal speed in a microcentrifuge to remove insoluble proteins. Then, the biotin-labeled proteins were pulled down with PuroMAG™ Magnetic Beads-Streptavidin at 100 μL of flurry beads/1 mL protein solution, rotate the sample tubes overnight at 4 °C. Then, the beads were washed twice with 1 mL of biotin-streptavidin lysis buffer with vigorous vortexing for 5 min followed by two washes with 2% SDS-PBS and then two washes with biotin-streptavidin lysis buffer. The proteins were eluted with two elution with 60 μL of 20 mM biotin in 50 mM Tris pH 8.3 and the identity of proteins were determined by high-quality and robust protein quantification by IonStar/nano-LC/MS at proteomics core facility run by Dr. Jun Qu [45] at the New York State Center of Excellence Bioinformatics and Life Sciences (CBLS), University at Buffalo.

3.2.5. Pathway Analysis of SC-62-16/SC-62-16Me Covalently Bound Protein

Genes encoding the proteins identified by mass spectrometry in SC-62-16 samples were used to run pathway enrichment analysis using g:Profiler [46] and functional annotation using DAVID [47,48]. For the pathway enrichment runs, KEGG [35] and Reactome database [36] were used for the pathway search. The pathways with Benjamini–Hochberg adjusted p-value < 0.05 were ranked by their p values with the lowest as the top hit.

Supplementary Materials: The following supporting information can be downloaded at: https://www.mdpi.com/article/10.3390/molecules30132696/s1, References [49–52] are cited in the supplementary materials.

Author Contributions: Conceptualization, X.W. and S.R.C.; methodology, X.W., R.L., J.Q., J.W. and S.R.C.; validation, X.W., S.R.C., J.W. and J.Q.; investigation, X.W., R.L., A.D.K., Y.Z. and C.X.; chemical synthesis and compound characterization, E.C.R., A.D.K. and S.L.G.; biological experiments, X.W. and R.L.; writing—original draft preparation, X.W. and S.R.C.; writing—review and editing, X.W., R.L., and S.R.C.; supervision, X.W. and S.R.C.; project administration, X.W. and S.R.C.; funding acquisition, X.W. and S.R.C. All authors have read and agreed to the published version of the manuscript.

Funding: This research was funded by Roswell Park Alliance Foundation to XW and SRC. This work was also supported by the NATIONAL CANCER INSITUTE (NCI), P30CA016056 involving the use of Roswell Park Comprehensive Cancer Center's shared resources. SRC thanks the National Institutes of Health for support of the compound synthesis (R35 GM153290 and R01 GM078383).

Institutional Review Board Statement: Not applicable.

Informed Consent Statement: Not applicable.

Data Availability Statement: All data are included in the article and Supplementary Materials.

Acknowledgments: We thank Yuping Wang for technical assistance in WB analysis.

Conflicts of Interest: The authors declare no conflicts of interest.

References

1. Zhong, L.; Li, Y.; Xiong, L.; Wang, W.; Wu, M.; Yuan, T.; Yang, W.; Tian, C.; Miao, Z.; Wang, T.; et al. Small molecules in targeted cancer therapy: Advances, challenges, and future perspectives. *Signal Transduct. Target. Ther.* **2021**, *6*, 201. [CrossRef]
2. Garg, P.; Malhotra, J.; Kulkarni, P.; Horne, D.; Salgia, R.; Singhal, S.S. Emerging Therapeutic Strategies to Overcome Drug Resistance in Cancer Cells. *Cancers* **2024**, *16*, 2478. [CrossRef] [PubMed]
3. Huang, L.; Jiang, S.; Shi, Y. Tyrosine kinase inhibitors for solid tumors in the past 20 years (2001–2020). *J. Hematol. Oncol.* **2020**, *13*, 143. [CrossRef] [PubMed]
4. Proietti, I.; Skroza, N.; Bernardini, N.; Tolino, E.; Balduzzi, V.; Marchesiello, A.; Michelini, S.; Volpe, S.; Mambrin, A.; Mangino, G.; et al. Mechanisms of Acquired BRAF Inhibitor Resistance in Melanoma: A Systematic Review. *Cancers* **2020**, *12*, 2801. [CrossRef]
5. Jung, J.; Lee, J.S.; Dickson, M.A.; Schwartz, G.K.; Le Cesne, A.; Varga, A.; Bahleda, R.; Wagner, A.J.; Choy, E.; de Jonge, M.J.; et al. TP53 mutations emerge with HDM2 inhibitor SAR405838 treatment in de-differentiated liposarcoma. *Nat. Commun.* **2016**, *7*, 12609. [CrossRef]
6. Chapeau, E.A.; Gembarska, A.; Durand, E.Y.; Mandon, E.; Estadieu, C.; Romanet, V.; Wiesmann, M.; Tiedt, R.; Lehar, J.; de Weck, A.; et al. Resistance mechanisms to TP53-MDM2 inhibition identified by in vivo piggyBac transposon mutagenesis screen in an Arf(-/-) mouse model. *Proc. Natl. Acad. Sci. USA* **2017**, *114*, 3151–3156. [CrossRef] [PubMed]
7. Sanz, G.; Singh, M.; Peuget, S.; Selivanova, G. Inhibition of p53 inhibitors: Progress, challenges and perspectives. *J. Mol. Cell Biol.* **2019**, *11*, 586–599. [CrossRef]
8. Arnoff, T.E.; El-Deiry, W.S. MDM2/MDM4 amplification and CDKN2A deletion in metastatic melanoma and glioblastoma multiforme may have implications for targeted therapeutics and immunotherapy. *Am. J. Cancer Res.* **2022**, *12*, 2102–2117.
9. Elmarakeby, H.A.; Hwang, J.; Arafeh, R.; Crowdis, J.; Gang, S.; Liu, D.; AlDubayan, S.H.; Salari, K.; Kregel, S.; Richter, C.; et al. Biologically informed deep neural network for prostate cancer discovery. *Nature* **2021**, *598*, 348–352. [CrossRef]
10. Hullein, J.; Slabicki, M.; Rosolowski, M.; Jethwa, A.; Habringer, S.; Tomska, K.; Kurilov, R.; Lu, J.; Scheinost, S.; Wagener, R.; et al. MDM4 Is Targeted by 1q Gain and Drives Disease in Burkitt Lymphoma. *Cancer Res.* **2019**, *79*, 3125–3138. [CrossRef]

11. Chen, J. The Cell-Cycle Arrest and Apoptotic Functions of p53 in Tumor Initiation and Progression. *Cold Spring Harb. Perspect. Med.* **2016**, *6*, a026104. [CrossRef]
12. Haronikova, L.; Bonczek, O.; Zatloukalova, P.; Kokas-Zavadil, F.; Kucerikova, M.; Coates, P.J.; Fahraeus, R.; Vojtesek, B. Resistance mechanisms to inhibitors of p53-MDM2 interactions in cancer therapy: Can we overcome them? *Cell. Mol. Biol. Lett.* **2021**, *26*, 53. [CrossRef]
13. Zhu, H.; Gao, H.; Ji, Y.; Zhou, Q.; Du, Z.; Tian, L.; Jiang, Y.; Yao, K.; Zhou, Z. Targeting p53-MDM2 interaction by small-molecule inhibitors: Learning from MDM2 inhibitors in clinical trials. *J. Hematol. Oncol.* **2022**, *15*, 91. [CrossRef] [PubMed]
14. Wu, W.; Xu, C.; Ling, X.; Fan, C.; Buckley, B.P.; Chernov, M.V.; Ellis, L.; Li, F.; Munoz, I.G.; Wang, X. Targeting RING domains of Mdm2-MdmX E3 complex activates apoptotic arm of the p53 pathway in leukemia/lymphoma cells. *Cell Death Dis.* **2015**, *6*, e2035. [CrossRef] [PubMed]
15. Itahana, K.; Mao, H.; Jin, A.; Itahana, Y.; Clegg, H.V.; Lindstrom, M.S.; Bhat, K.P.; Godfrey, V.L.; Evan, G.I.; Zhang, Y. Targeted inactivation of Mdm2 RING finger E3 ubiquitin ligase activity in the mouse reveals mechanistic insights into p53 regulation. *Cancer Cell* **2007**, *12*, 355–366. [CrossRef]
16. Pant, V.; Xiong, S.; Iwakuma, T.; Quintas-Cardama, A.; Lozano, G. Heterodimerization of Mdm2 and Mdm4 is critical for regulating p53 activity during embryogenesis but dispensable for p53 and Mdm2 stability. *Proc. Natl. Acad. Sci. USA* **2011**, *108*, 11995–12000. [CrossRef] [PubMed]
17. Huang, L.; Yan, Z.; Liao, X.; Li, Y.; Yang, J.; Wang, Z.G.; Zuo, Y.; Kawai, H.; Shadfan, M.; Ganapathy, S.; et al. The p53 inhibitors MDM2/MDMX complex is required for control of p53 activity in vivo. *Proc. Natl. Acad. Sci. USA* **2011**, *108*, 12001–12006. [CrossRef] [PubMed]
18. Chinnam, M.; Xu, C.; Lama, R.; Zhang, X.; Cedeno, C.D.; Wang, Y.; Stablewski, A.B.; Goodrich, D.W.; Wang, X. MDM2 E3 ligase activity is essential for p53 regulation and cell cycle integrity. *PLoS Genet.* **2022**, *18*, e1010171. [CrossRef]
19. Lama, R.; Xu, C.; Galster, S.L.; Querol-Garcia, J.; Portwood, S.; Mavis, C.K.; Ruiz, F.M.; Martin, D.; Wu, J.; Giorgi, M.C.; et al. Small molecule MMRi62 targets MDM4 for degradation and induces leukemic cell apoptosis regardless of p53 status. *Front. Oncol.* **2022**, *12*, 933446. [CrossRef]
20. Li, J.; Lama, R.; Galster, S.L.; Inigo, J.R.; Wu, J.; Chandra, D.; Chemler, S.R.; Wang, X. Small-Molecule MMRi62 Induces Ferroptosis and Inhibits Metastasis in Pancreatic Cancer via Degradation of Ferritin Heavy Chain and Mutant p53. *Mol. Cancer Ther.* **2022**, *21*, 535–545. [CrossRef]
21. Rew, Y.; Sun, D. Discovery of a small molecule MDM2 inhibitor (AMG 232) for treating cancer. *J. Med. Chem.* **2014**, *57*, 6332–6341. [CrossRef] [PubMed]
22. Lama, R.; Galster, S.L.; Xu, C.; Davison, L.W.; Chemler, S.R.; Wang, X. Dual Targeting of MDM4 and FTH1 by MMRi71 for Induced Protein Degradation and p53-Independent Apoptosis in Leukemia Cells. *Molecules* **2022**, *27*, 7665. [CrossRef]
23. Shultz, M.D. Two Decades under the Influence of the Rule of Five and the Changing Properties of Approved Oral Drugs. *J. Med. Chem.* **2019**, *62*, 1701–1714. [CrossRef] [PubMed]
24. Fleming, F.F.; Yao, L.; Ravikumar, P.C.; Funk, L.; Shook, B.C. Nitrile-containing pharmaceuticals: Efficacious roles of the nitrile pharmacophore. *J. Med. Chem.* **2010**, *53*, 7902–7917. [CrossRef] [PubMed]
25. Sosman, J.A.; Kim, K.B.; Schuchter, L.; Gonzalez, R.; Pavlick, A.C.; Weber, J.S.; McArthur, G.A.; Hutson, T.E.; Moschos, S.J.; Flaherty, K.T.; et al. Survival in BRAF V600-mutant advanced melanoma treated with vemurafenib. *N. Engl. J. Med.* **2012**, *366*, 707–714. [CrossRef]
26. Bollag, G.; Hirth, P.; Tsai, J.; Zhang, J.; Ibrahim, P.N.; Cho, H.; Spevak, W.; Zhang, C.; Zhang, Y.; Habets, G.; et al. Clinical efficacy of a RAF inhibitor needs broad target blockade in BRAF-mutant melanoma. *Nature* **2010**, *467*, 596–599. [CrossRef]
27. McLean, L.R.; Zhang, Y.; Li, H.; Li, Z.; Lukasczyk, U.; Choi, Y.M.; Han, Z.; Prisco, J.; Fordham, J.; Tsay, J.T.; et al. Discovery of covalent inhibitors for MIF tautomerase via cocrystal structures with phantom hits from virtual screening. *Bioorg. Med. Chem. Lett.* **2009**, *19*, 6717–6720. [CrossRef]
28. Shergalis, A.; Xue, D.; Gharbia, F.Z.; Driks, H.; Shrestha, B.; Tanweer, A.; Cromer, K.; Ljungman, M.; Neamati, N. Characterization of Aminobenzylphenols as Protein Disulfide Isomerase Inhibitors in Glioblastoma Cell Lines. *J. Med. Chem.* **2020**, *63*, 10263–10286. [CrossRef]
29. Sutanto, F.; Konstantinidou, M.; Domling, A. Covalent inhibitors: A rational approach to drug discovery. *RSC Med. Chem.* **2020**, *11*, 876–884. [CrossRef]
30. Dufrasne, F.; Gelbcke, M.; Neve, J.; Kiss, R.; Kraus, J.L. Quinone methides and their prodrugs: A subtle equilibrium between cancer promotion, prevention, and cure. *Curr. Med. Chem.* **2011**, *18*, 3995–4011. [CrossRef]
31. Boike, L.; Henning, N.J.; Nomura, D.K. Advances in covalent drug discovery. *Nat. Rev. Drug Discov.* **2022**, *21*, 881–898. [CrossRef] [PubMed]

32. Wright, M.H.; Sieber, S.A. Chemical proteomics approaches for identifying the cellular targets of natural products. *Nat. Prod. Rep.* **2016**, *33*, 681–708. [CrossRef]
33. Lapinsky, D.J. Tandem photoaffinity labeling-bioorthogonal conjugation in medicinal chemistry. *Bioorg. Med. Chem.* **2012**, *20*, 6237–6247. [CrossRef] [PubMed]
34. Kolb, H.C.; Sharpless, K.B. The growing impact of click chemistry on drug discovery. *Drug Discov. Today* **2003**, *8*, 1128–1137. [CrossRef] [PubMed]
35. Kanehisa, M.; Goto, S. KEGG: Kyoto encyclopedia of genes and genomes. *Nucleic Acids Res.* **2000**, *28*, 27–30. [CrossRef]
36. Joshi-Tope, G.; Gillespie, M.; Vastrik, I.; D'Eustachio, P.; Schmidt, E.; de Bono, B.; Jassal, B.; Gopinath, G.R.; Wu, G.R.; Matthews, L.; et al. Reactome: A knowledgebase of biological pathways. *Nucleic Acids Res.* **2005**, *33*, D428–D432. [CrossRef]
37. Kerek, E.M.; Cromwell, C.R.; Hubbard, B.P. Identification of Drug Resistance Genes Using a Pooled Lentiviral CRISPR/Cas9 Screening Approach. *Methods Mol. Biol.* **2021**, *2381*, 227–242. [CrossRef]
38. Mencher, S.K.; Wang, L.G. Promiscuous drugs compared to selective drugs (promiscuity can be a virtue). *BMC Clin. Pharmacol.* **2005**, *5*, 3. [CrossRef]
39. Shen, Y.; Radhakrishnan, M.L.; Tidor, B. Molecular mechanisms and design principles for promiscuous inhibitors to avoid drug resistance: Lessons learned from HIV-1 protease inhibition. *Proteins* **2015**, *83*, 351–372. [CrossRef]
40. Korikani, M.; Fathima, N.; Nadiminti, G.; Akula, S.; Kancha, R.K. Applications of promiscuity of FDA-approved kinase inhibitors in drug repositioning and toxicity. *Toxicol. Appl. Pharmacol.* **2023**, *465*, 116469. [CrossRef]
41. Godar, S.; Ince, T.A.; Bell, G.W.; Feldser, D.; Donaher, J.L.; Bergh, J.; Liu, A.; Miu, K.; Watnick, R.S.; Reinhardt, F.; et al. Growth-inhibitory and tumor- suppressive functions of p53 depend on its repression of CD44 expression. *Cell* **2008**, *134*, 62–73. [CrossRef] [PubMed]
42. Chou, T.C. Theoretical basis, experimental design, and computerized simulation of synergism and antagonism in drug combination studies. *Pharmacol. Rev.* **2006**, *58*, 621–681. [CrossRef] [PubMed]
43. Pradhan, A.J.; Lu, D.; Parisi, L.R.; Shen, S.; Berhane, I.A.; Galster, S.L.; Bynum, K.; Monje-Galvan, V.; Gokcumen, O.; Chemler, S.R.; et al. Protein acylation by saturated very long chain fatty acids and endocytosis are involved in necroptosis. *Cell Chem. Biol.* **2021**, *28*, 1298–1309.e7. [CrossRef]
44. Cheah, J.S.; Yamada, S. A simple elution strategy for biotinylated proteins bound to streptavidin conjugated beads using excess biotin and heat. *Biochem. Biophys. Res. Commun.* **2017**, *493*, 1522–1527. [CrossRef]
45. Shen, S.; Wang, X.; Zhu, X.; Rasam, S.; Ma, M.; Huo, S.; Qian, S.; Zhang, M.; Qu, M.; Hu, C.; et al. High-quality and robust protein quantification in large clinical/pharmaceutical cohorts with IonStar proteomics investigation. *Nat. Protoc.* **2023**, *18*, 700–731. [CrossRef] [PubMed]
46. Raudvere, U.; Kolberg, L.; Kuzmin, I.; Arak, T.; Adler, P.; Peterson, H.; Vilo, J. g:Profiler: A web server for functional enrichment analysis and conversions of gene lists (2019 update). *Nucleic Acids Res.* **2019**, *47*, W191–W198. [CrossRef]
47. Sherman, B.T.; Hao, M.; Qiu, J.; Jiao, X.; Baseler, M.W.; Lane, H.C.; Imamichi, T.; Chang, W. DAVID: A web server for functional enrichment analysis and functional annotation of gene lists (2021 update). *Nucleic Acids Res.* **2022**, *50*, W216–W221. [CrossRef]
48. Huang, D.W.; Sherman, B.T.; Lempicki, R.A. Systematic and integrative analysis of large gene lists using DAVID bioinformatics resources. *Nat. Protoc.* **2009**, *4*, 44–57. [CrossRef]
49. Zhang, M.; Gallagher, J.A.; Coppock, M.B.; Pantzar, L.M.; Williams, M.E. Cooperative Assembly of Zn Cross-Linked Artificial Tripeptides with Pendant Hydroxyquinoline Ligands. *Inorg. Chem.* **2012**, *51*, 11315–11323. [CrossRef]
50. Warner, V.D.; Sane, J.N.; Mirth, D.B. Synthesis and in Vitro Evaluation of 8-Hydroxyquinoline Analogs as Inhibitors of Dental Plaque. *J. Med. Chem.* **1976**, *19*, 167–169. [CrossRef]
51. Xiao, C.; Cheng, Y.; Zhang, Y.; Ding, J.; He, C.; Zhuang, X.; Chen, X. Side Chain Impacts on pH- and Thermo-Responsiveness of Tertiary Amine Functionalized Polypeptides. *J. Polym. Sci. Part A* **2014**, *52*, 671–679. [CrossRef]
52. Malerich, J.P.; Hagihara, K.; Rawal, V.H. Chiral Squaramide Derivatives are Excellent Hydrogen Bond Donor Catalysts. *J. Am. Chem. Soc.* **2008**, *130*, 14416–14417. [CrossRef] [PubMed]

molecules

MDPI

Article

Structural Insights into Salinosporamide a Mediated Inhibition of the Human 20S Proteasome

Hagen Sülzen [1], Pavla Fajtova [1,2], Anthony J. O'Donoghue [2,3], Jan Silhan [1,*] and Evzen Boura [1,*]

1 Institute of Organic Chemistry and Biochemistry, Czech Academy of Sciences, Flemingovo namesti 2, 16610 Prague, Czech Republic; fajtova@uochb.cas.cz (P.F.)
2 Skaggs School of Pharmacy and Pharmaceutical Sciences, University of California San Diego, La Jolla, CA 92093, USA; ajodonoghue@health.ucsd.edu
3 Center for Discovery and Innovation in Parasitic Diseases, Skaggs School of Pharmacy and Pharmaceutical Sciences, University of California San Diego, La Jolla, CA 92093, USA
* Correspondence: jan.silhan@uochb.cas.cz (J.S.); evzen.boura@uochb.cas.cz (E.B.)

Abstract: The 20S proteasome, a critical component of the ubiquitin–proteasome system, plays a central role in regulating protein degradation in eukaryotic cells. Marizomib (MZB), also known as salinosporamide A, is a natural γ-lactam-β-lactone compound derived from *Salinispora tropica* and is a potent 20S proteasome covalent inhibitor with demonstrated anticancer properties. Its broad-spectrum inhibition of all three proteasome subunits and its ability to cross the blood–brain barrier has made it a promising therapeutic candidate for glioblastoma. In addition to this, MZB also demonstrates significant inhibition against the 20S proteasome of *Trichomonas vaginalis* (*Tv*20S), a protozoan parasite, suggesting its potential for parasitic treatments. Here, we present the cryo-EM structure of the human 20S proteasome in complex with MZB at 2.55 Å resolution. This structure reveals the binding mode of MZB to all six catalytic subunits within the two β-rings of the 20S proteasome, providing a detailed molecular understanding of its irreversible inhibitory mechanism. These findings enhance the therapeutic potential of MZB for both cancer and parasitic diseases at the molecular level and highlight marine-derived natural products in targeting the proteasome for therapeutic applications.

Keywords: 20S; proteasome; marizomib; MZB; cryo-EM

1. Introduction

Marizomib (MZB), also known as salinosporamide A, is a natural γ-lactam-β-lactone compound isolated from the marine bacterium *Salinispora tropica* [1]. It is a non-peptidic 20S proteasome covalent inhibitor [1,2] that has garnered significant attention for its potent anticancer properties and blood–brain barrier permeability [3–5]. This unique capability makes marizomib particularly promising for the treatment of glioblastoma [3] and other central nervous system malignancies.

The proteasome is a critical component of the ubiquitin–proteasome system, which regulates protein degradation in eukaryotic cells [6,7]. Structurally, the proteasome adopts a barrel-like architecture consisting of four heptameric rings, each formed by seven α or β subunits [8–10]. The inner two rings harbour six catalytically active subunits (β1, β2 and β5) that possess N-terminal threonine residues critical for their enzymatic function [11]. Despite the conserved architecture of these active sites, the substrate binding pockets preferentially cleave on the C-terminal side of hydrophobic amino acids (β5), positively charged residues (β2) or negatively charged residues (β1). These variations confer chymotrypsin-like, trypsin-like and caspase-like substrate specificity to the proteasome, respectively [12].

As proteasome activity is critical to numerous cellular processes, its dysfunction is associated with a variety of diseases [13,14]. Cancer cells, characterized by rapid growth and genetic instability, are particularly dependent on proteasome activity to manage the large quantities of aberrant proteins they produce. Proteasome inhibitors like MZB disrupt this process, leading to the accumulation of misfolded proteins, cellular stress and ultimately apoptosis. This mechanism of action is especially effective in cancer cells, which are more reliant on proteasome function than normal cells.

Marizomib irreversibly binds to the catalytic threonine of the 20S proteasome (Figure 1). As it lacks a peptide moiety, MZB does not bind in a substrate-like manner and therefore can target all three subunits [15–17]. MZB has highest affinity to the β5 subunit but readily binds to the β1 and β2 subunits at higher concentrations [18,19]. This broad inhibition profile is likely to contribute to its potent anticancer activity, even in tumours resistant to other proteasome inhibitors. Furthermore, the unique β-lactone pharmacophore is highly specific for reacting with the catalytic threonine residues, as no other human hydrolases have been shown to be targeted by MZB. In addition to its anti-cancer effects, MZB is being investigated for applications in infectious diseases, such as malaria, by targeting the proteasome of *Plasmodium falciparum* [20].

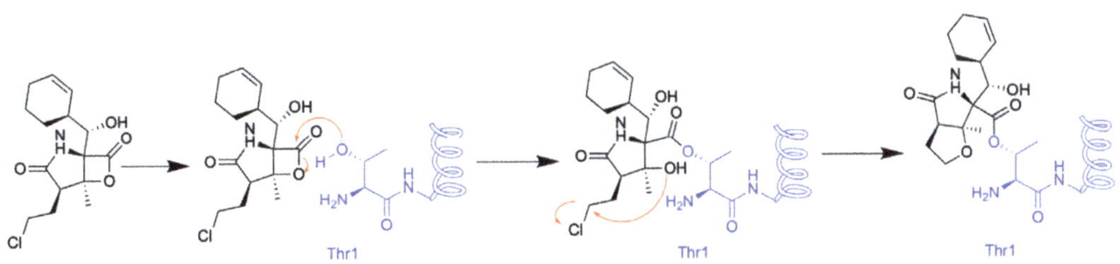

Figure 1. Schematic of the putative reaction mechanism between MZB and the *N*-terminal threonine of proteasome subunits.

Marizomib exemplifies the therapeutic potential of oceanic biodiversity in drug discovery. Beyond its ability to cross the blood–brain barrier, it also has enhanced stability and bioavailability compared to other peptide-like proteasome inhibitors. Currently in Phase III clinical trials, MZB has been evaluated in combination with standard glioblastoma treatments, such as temozolomide-based chemoradiotherapy. Recent trial results revealed that there was no significant improvement in overall survival or progression-free survival in patients with newly diagnosed glioblastoma. In addition, patients receiving MZB experienced more documented adverse events when compared to those receiving the standard temozolomide-based chemoradiotherapy [3]. However, ongoing research continues to explore the use of MZB in combination therapies to enhance efficacy and address resistance mechanisms (ClinicalTrials.gov: NCT02330562).

The molecular details of how MZB binds to all three catalytic β-subunits have not been described. Therefore, in this study, we used single-particle cryo-EM to provide a detailed molecular explanation of the binding interactions with the human 20S proteasome.

2. Results

2.1. Biochemical Validation of the Human 20S Proteasome

The commercially obtained human 20S proteasome (h20S) was evaluated in biochemical assays for catalytic activity using fluorogenic reporter substrates that are each specific for either β1, β2 or β5 (Figure 2). After confirming the activity of all three catalytic subunits, we assessed whether these activities could be inhibited in the presence of MZB. Using a concentration range of 0 to 12.5 μM of MZB, we could demonstrate that the β5 subunits were inhibited with an IC_{50} of 18.5 nM while the IC_{50} for the β2 and β1 were 326.5 nM and 596.6 nM, respectively (Figure 2). Under these conditions all subunits were completely inhibited with 12.5 μM of MZB. These studies validate the quality of both the enzyme and inhibitor for structural studies.

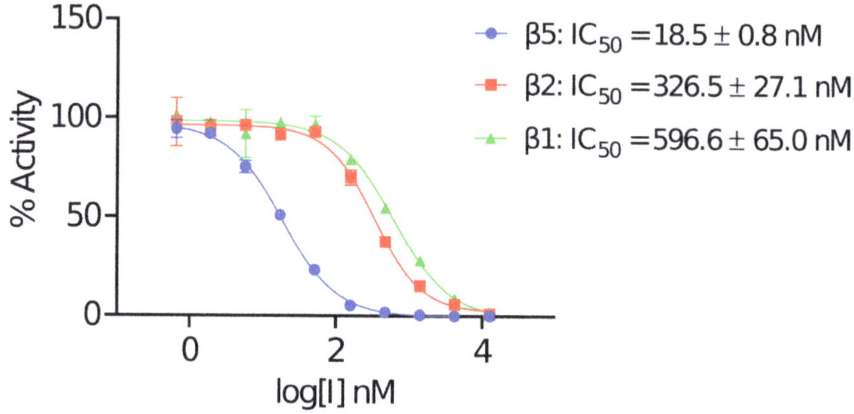

Figure 2. Half-maximal inhibitory concentration (IC_{50}) of MZB for the proteolytic h20S subunits. IC_{50} curves for the inhibition of individual h20S proteasome subunits (β1, β2 and β5) determined using fluorogenic substrates Z-LLE-amc, Z-VLR-amc and Suc-LLVY-amc. Assays were performed in 3 technical replicates, and data are presented as mean values ± SD.

2.2. Structural Characterisation of the h20S Proteasome

To gain structural insights into MZB-mediated inhibition of the h20S, we employed single particle cryo-EM (Figure S1). The complex could be reconstructed to a reported $FSC_{0.143}$ of 2.55 Å (Figures 3A and S2). Some of the human 20S particles formed filamentous structures upon vitrification on cryo-EM grids, which we assumed to be an artifact of the cryo-EM freezing process and had no visual effect on cryo-EM density (Figure S1). While most particles observed in the micrographs appeared to present either-side or top-and-bottom views of the complex, continuous coverage of the Euler angle distribution allowed for a complete reconstruction (Figure S2B). The local resolution of the reconstruction ranges predominantly between 2.53 and 5.7 Å (25th to 75th percentile), with lower resolutions observed primarily in regions corresponding to the solvent-facing surfaces of the α-subunits (Figure S2C). The h20S proteasome could be modelled in its entirety except for flexible regions, $\alpha1_{\text{Ala246-Asp249}}$, $\alpha2_{\text{Ala235-Ala237}}$, $\alpha3_{\text{Gln257-Lys262}}$, $\alpha4_{\text{Lys61}}$, $\alpha4_{\text{Glu241-Ser252}}$, $\alpha5_{\text{Val242-Ile246}}$, $\alpha6_{\text{Glu241-His266}}$, $\alpha7_{\text{Glu246-Met259}}$, $\alpha1_{\text{Pro203-Ala204}}$, $\beta2_{\text{Val222-Ser233}}$, $\beta4_{\text{Lys198-Ser201}}$, $\beta5_{\text{Ser201-Pro203}}$, $\beta6_{\text{Asp242}}$ and $\beta\ 6_{\text{Gly262-Glu264}}$, covering 6256 of 6452 residues (96.96%) in total. An average map-to-model correlation of 0.92 ($CC_{\text{side chain}}$, calculated with phenix.validation_cryoem [21]) indicates an excellent fit of the model to the experimental data (Figure S2D).

As expected, the atomic model obtained for the h20S proteasome in complex with MZB closely resembles the characteristic architecture commonly shared amongst 20S proteasomes and displays C2 symmetry, with the two sets of 14 subunits arranged in the conventional ring configuration of α1–α7, β1–β7/β1–β7, α1–α7. Unsurprisingly, the h20S/MZB model could be aligned to the h20S structure used as a starting model (PDB: 7PG9) with an RMSD of 0.97 (27449 atoms), confirming an overall nearly identical fold.

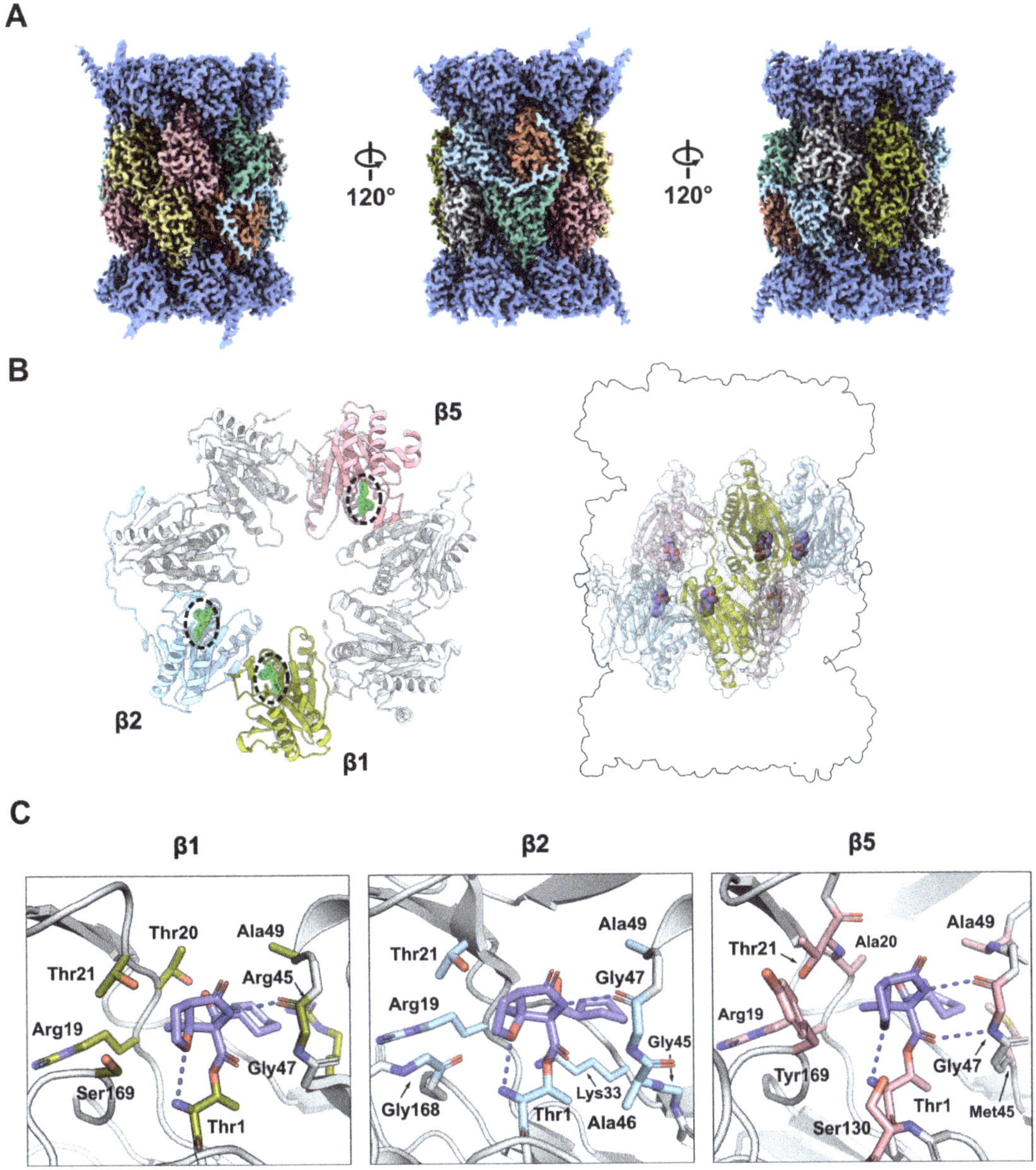

Figure 3. Cryo-EM structure of h20S in complex with MZB. (**A**) Final cryo-EM reconstruction after local sharpening using EMReady [22]. Density corresponding to the α-subunits is coloured in dark blue, proteolytic subunits β1, β2 and β5 are coloured in olive, cyan and pink, respectively. The remaining subunits β3, β4, β6 and β7 are coloured in orange, yellow, mint and grey respectively. (**B**) Cartoon representation of one β subunit ring (left) and placement of the catalytic subunits relative to the overall proteasome architecture (right). The catalytic subunits follow the same colour scheme as in (**A**), the remaining subunits are coloured in light grey. The cryo-EM density for MZB is shown as green mesh (left), and the atomic model is depicted as spheres (right). (**C**) Close-up views of the three distinct active sites in β1, β2 and β5. MZB and interacting residues are displayed as coloured sticks. The colour scheme is consistent with (**B**). Side chains of non-interacting residues (white) have been removed for clarity. Hydrogen bonds are shown as blue dashed lines.

2.3. Binding of MZB to Catalytic Subunits B1, B2 and B5

The human 20S proteasome contains three proteolytically active subunits, β1, β2 and β5, within each β-subunit ring. Each active site features a conserved catalytic triad comprised of Thr1, Asp17 and Lys33, in addition to residues Ser129, Asp166 and Ser169, which are thought to provide structural integrity to the proteolytic centre and enhance catalysis [23] The well-defined density of the sharpened cryo-EM reconstruction allowed for unambiguous placement of the atomic model of the inhibitor (CC_{Ligand} = 0.93) (Figures 3B and S3). The local resolution within a 10 Å radius of the catalytic sites is 2.34–2.75 Å for β1, 2.34–2.72 Å for β2 and 2.34–2.62 Å for β5, supporting our confident placement of MZB.

As expected, the carbonyl carbon atom derived from the beta-lactone ring of MZB is covalently bound to the oxygen from the hydroxyl side-chain group of the N-terminal catalytic Thr1 residue in all proteolytic subunits. Similarly, the orientation of the inhibitor is consistent across all three catalytic sites with the cyclohexenyl moiety occupying the so-called S1 pocket. While few hydrophobic interactions stabilize the ring structure in the S1 pocket, MZB is primarily coordinated via hydrogen bonding (Figure 3C). In all three active sites, residues Thr1 and Gly47 appear to coordinate the inhibitor via main-chain interactions, while for subunits β1 and β5, the cryo-EM reconstruction additionally places the peptide nitrogen (and carbonyl group) of the residue Thr21 in coordination distance of the free carboxyl group in MZB (Figure 3C). Lastly, residue Ser130 in the catalytic site of the β5 subunit appears sufficiently close to coordinate the nitrogen of the γ-lactam ring of MZB via a sidechain interaction (Figure 3C). Intriguingly, the increased number of coordination sites for MZB in the active site of the β5 subunit correlates well with the higher potency of the inhibitor to this subunit (Figure 2). More detailed descriptions of the interactions between MZB and the active site pockets are shown in Figure S4.

2.4. Comparison of MZB Binding to the Active Proteasome Sites of Human and T. vaginalis

MZB is a pan-proteasome inhibitor and has been suggested as a potential treatment for diseases caused by eukaryotic parasites. Recently, we determined the structure of the *Trichomonas vaginalis* 20S proteasome in complex with MZB [19]. We compared how MZB binds these two distinct proteasomes, as these differences could be significant for future drug design.

While both, the human and *T. vaginalis* 20S proteasome (*Tv*20S), generally exhibit the commonly shared and characteristic architecture, structural alignment across all atoms reveals a root mean square deviation (RMSD) of 3.184 Å (Figure 4A), a surprising discrepancy in light of the rather high conservation of the individual subunits, ranging from 25–51% sequence identity [19,24]. The largest differences between the atomic models are observed in the α-subunit ring (Figure 4A). *Tv*20S sample remained largely monodisperse [19]. Alternatively, differences in the overall packing of the individual subunits could result in structural differences reflected by the inflated RMSD. Individual structural alignment of the proteolytic subunits results in RMSD (β1 = 0.865 Å, β2 = 0.782 Å and β5 = 0.820 Å), the values between 0.78 and 0.87 Å accentuate the conserved nature of the active sites.

Indeed, structural comparison of the β1 and β2 subunits reveal nearly identical orientation and coordination of MZB (Figure 4B). While the residue Thr21 of the h20 β1 subunit, predicted to form a hydrogen bond with MZB (Figure 4C), is replaced by Ser21 in the *T. vaginalis* equivalent, the peptide nitrogen is within bonding distance to the free carboxyl group of the inhibitor as well (3.7 Å).

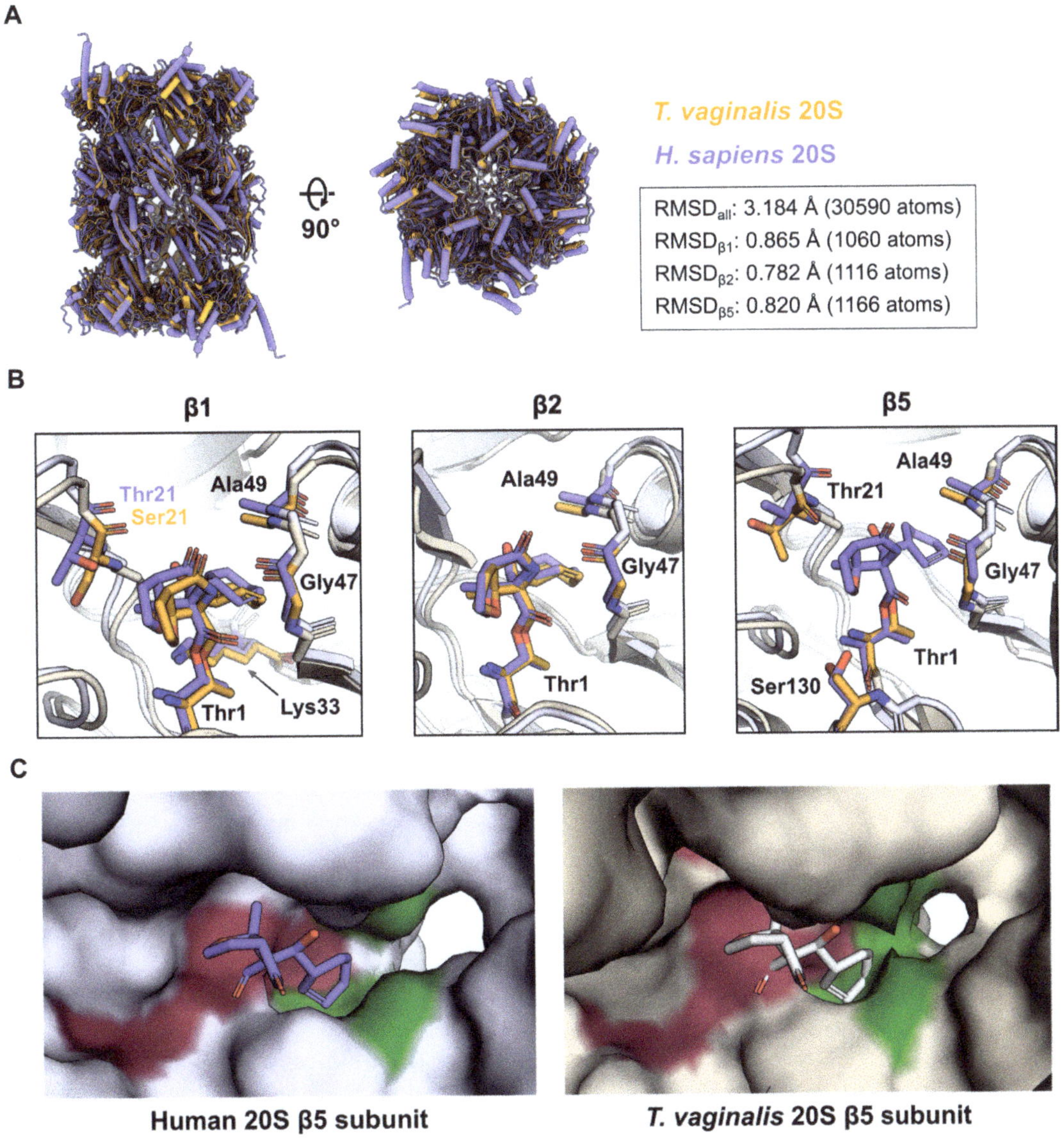

Figure 4. Comparison of h20S and *Tv*20S in complex with MZB. (**A**) Alignment and structural comparison of human (purple, PDB: 9HMN) and *T. vaginalis* (orange, PDB: 8OIX) 20S proteasomes bound to MZB. RMSD values for the alignment of the whole molecule and the individual catalytic subunits were obtained via PyMOL align with 5 cycles of outlier rejection. (**B**) Close-up views of the β1, β2 and β5 active sites. MZB and interacting residues are displayed as coloured sticks. The colour scheme is consistent with (**A**). Side chains of non-interacting residues (light purple/orange) have been removed for clarity. Labels for identical amino acids across both complexes are shown in black, deviating amino acids are shown in their respective colour. (**C**) Surface representation of the β5 subunit of human (left) and *T. vaginalis* (right) 20S proteasome. The catalytic site and S1 pocket are coloured pink and green, respectively. The atomic model for the *Tv*20S β5 subunit (PDB: 8OIX) does not include a model of MZB. Instead, MZB bound to the human protein has been aligned, superimposed and is shown as transparent sticks for representative purposes.

Intriguingly, our previous cryo-EM reconstruction of *Tv*20S had insufficient density to include MZB in the final model of the β5 subunit, despite the fact that the inhibitor attenuated proteolytic activity almost entirely in vitro [19]. Alignment of the active site to the human β5 subunit once more reveals a nearly identical arrangement (Figure 4B), indicating structural differences elsewhere. As described earlier, in all three proteolytic subunits of h20S, the cyclohexenyl moiety of the inhibitor occupies the S1 pocket. When comparing the overall structural arrangement of the β5 subunits, it is apparent that access to the S1 cavity is much more restricted in *T. vaginalis* (Figure 4C). It is conceivable that, because of this steric impairment, a reduced number of coordination sites are available, granting the covalently bound MZB increased flexibility and thereby smearing its cryo-EM density. It is noteworthy that MZB inhibits all three active β subunits of *Tv*20S in a manner similar to human 20S, with IC_{50} values of β1 = 243.8 ± 72.2 nM, β2 = 283.7 ± 31.0 nM and β5 = 3.9 ± 0.2 nM, respectively (Figure S5).

3. Discussion

Proteasomal degradation of ubiquitinated proteins is a key cellular process that controls protein turnover and thus plays a crucial role in numerous cellular functions. These include development and differentiation, DNA repair, cell-cycle progression, immune responses, apoptosis and stress adaptation. The ability to selectively degrade specific targets is crucial to maintain cellular homeostasis and to dynamically regulate essential biological pathways [25].

Proteasome inhibition disrupts these tightly regulated processes by preventing protein degradation, which induces cellular stress and, eventually, leads to apoptosis. This mechanism is particularly effective in selectively killing rapidly dividing cancer cells such as myeloma cells. These cells rely heavily on protein integrity to sustain their high production of gamma globulin [26].

Consequently, FDA-approved proteasome inhibitors, such as bortezomib, carfilzomib and ixazomib, are used to treat multiple myeloma and other cancers [27,28]. New and promising drugs, such as marizomib (MZB), oprozomib and delanzomib, are currently undergoing testing, either as standalone treatments or in combination with other therapeutics [29].

Structural information is crucial for guiding the development of new proteasome inhibitors with enhanced efficacy and specificity. The knowledge of MZB's interactions with the human proteasome can help design novel therapeutics for various cancers. In this study, we investigated the binding of MZB to the human 20S proteasome using cryo-EM, revealing its molecular architecture and specific interactions. High-resolution cryo-EM maps confirmed that MZB irreversibly binds to all three catalytic subunits of the proteasome: chymotrypsin-like (β5), trypsin-like (β2) and caspase-like (β1). This pan-inhibitory activity distinguishes MZB from other proteasome inhibitors, such as bortezomib and carfilzomib [28]. Our findings provide structural insights at near-atomic resolution, explaining the broad inhibitory action of MZB. This knowledge can serve as a foundation for optimizing MZB's application or redesigning it to achieve more targeted therapeutic effects.

Furthermore, using a combination of proteasome inhibitors with other drugs may pave the way for safer and more specific cancer therapies. Preclinical studies have demonstrated promising results for MZB in triple-negative breast cancer (TNBC), where it effectively inhibits primary tumour growth and metastasis to the lungs and brain, potentially addressing the limitations of existing therapies like bortezomib and carfilzomib [30]. In cervical cancer, combining MZB with cisplatin shows synergistic effects both in vitro and in vivo, underscoring the need for further investigation into optimal dosing and scheduling [31].

Additionally, the potential of MZB combinations with other anti-cancer agents, such as glycolysis inhibitors, is particularly compelling, given the observed upregulation of glycolysis pathway proteins following MZB treatment in TNBC cells [30]. However, further research is required to evaluate the long-term efficacy of MZB, identify potential resistance mechanisms and refine combination strategies with other therapeutics. Understanding how MZB impacts cell-cycle progression, apoptosis and protein degradation pathways will be critical, particularly for designing synergistic approaches. Moreover, addressing whether prolonged MZB exposure induces acquired resistance, as observed with previous proteasome inhibitors, will be crucial for maximizing its therapeutic potential [28,32]. These efforts will be instrumental in optimizing MZB for the treatment of cancer and, potentially, to treat parasitic infections [33,34].

Beyond cancer treatments, MZB shows potential for combating parasitic infections. Our comparative analysis of MZB binding to the human 20S proteasome and the eukaryotic parasite *Trichomonas vaginalis* (*Tv*20S) proteasome reveals a promising pathway for developing targeted treatments against eukaryotic parasites. Structural insights into the binding of MZB to *Tv*20S suggest that it could be chemically modified to selectively target the *Tv*20S β5 subunit over the human β5 subunit. A previous study has demonstrated that MZB can be modified at the P2 position to enhance its reactivity and selectivity [34]. Specifically, substituting the original chloroethyl group with alternative functionalities such as bromoethyl, iodoethyl or tosyloxyethyl retains the leaving group character while improving reactivity. Conversely, incorporating non-leaving groups like thiocyanoethyl, azidoethyl, hydroxyethyl or ethyl significantly alters the electronic and physicochemical properties at this position [35]. These structural modifications influence molecular weight, lipophilicity and hydrogen-bonding capabilities, which could ultimately improve pharmacokinetic profiles, target specificity and the overall therapeutic index of these proteasome inhibitors. As a result, such modification could lead to drugs highly specific to *T. vaginalis*, offering novel therapeutic options for diseases caused by this parasite while minimizing off-target effects on human cells.

In conclusion, elucidating MZB's mechanism of action represents a critical step toward designing more effective therapeutics. MZB could be improved to more effectively treat cancer and, moreover, could be transformed into a drug to effectively target eukaryotic parasites such as *T. vaginalis*.

4. Materials and Methods

Human constitutive proteasome (h20S, #E-360, R&D Systems), marizomib (MZB, MedChemExpress), fluorogenic substrates Suc-LLVY-amc, Z-VLR-amc and Z-LLE-amc were purchased from Cayman Chemical, Ann Arbor, MI, USA. Substrates were dissolved in DMSO at a concentration of 10 mM and stored at −20 °C. MZB and h20S were aliquoted and stored at −80 °C.

4.1. Half-Maximal Inhibitory Concentration (IC_{50}) Assays

The inhibition of individual h20S subunits was confirmed using subunit-specific fluorogenic substrates for the β1, β2 and β5 subunits, namely Z-LLE-amc, Z-VLR-amc and Suc-LLVY-amc. Kinetic assays were performed in 384-well plates using 1 nM h20S preincubated with 100 nM PA28α human activator. The PA28α activator was expressed in *E. coli* using the pSumo vector, as previously described [33]. The assay was conducted in a reaction buffer containing 50 mM HEPES (pH 7.5) and 1 mM DTT, with substrate concentrations of 80 μM Z-LLE-amc, 30 μM Z-VLR-amc and 65 μM Ac-LLVY-amc, in a final volume of 8 μL per well. Inhibitors were dispensed into the plates using an Echo650 liquid handler (Beckman Coulter, Brea, CA, USA). Pre-steady-state kinetic measurements were carried out for the irreversible inhibitor, and IC_{50} values were calculated 60 to 90 min after initiating the reaction. Data analysis was performed using GraphPad Prism software version 10.4.1. All assays were conducted in triplicate using 384-well black plates (Greiner Bio-One, Monroe, NC, USA) at 37 °C. Fluorescence measurements were recorded with a Synergy HTX Multi-Mode Microplate Reader (BioTek, Winooski, VT, USA), using excitation and emission wavelengths of 360 nm and 460 nm, respectively.

4.2. Preparation of Cryo-EM Grids and Data Acquisition

h20S (approx. 0.6 μM) in a buffer consisting of 20 mM HEPES pH 7.5, 50 mM NaCl and 0.25 mM THP (Tris(hydroxypropyl)phosphine; Sigma-Aldrich, St. Louis, MO, USA) was incubated with 50 μM MZB at room temperature for 1 h before cooling the sample on ice. We applied 4 μL of the proteasome-inhibitor complex (0.45 mg/mL) to freshly glow discharged Quantifoil R2/1 300-mesh copper grids (EM Sciences Hatfield, PA, USA, Prod. No. Q350CR1). Excess sample was removed by blotting with an FEI Vitrobot Mark IV (Thermo Fisher Scientific, Waltham, MA, USA) (4 °C, 100% humidity, blot force −5) before plunge freezing the grids in liquid ethane.

The sample was imaged on a Titan Krios G3i microscope (Thermo Fisher Scientific, Waltham, MA, USA) equipped with a Gatan K3 detector and operated at 300 kV. We recorded 4625 multi-frame movies (40 frames with a total dose of 60 $e^-/Å^2$) at a magnification of 105,000×, yielding a final pixel size of 0.8336 Å. Data were collected using the EPU v 3.0.0 data collection software (Thermo Fisher Scientific, Waltham, MA, USA).

4.3. Cryo-EM Data Processing

All image processing steps were performed in cryoSPARC (version 4.6.2) [36]. Movies were motion and CTF corrected using the Patch Motion and Patch CTF (both cryoSPARC 4.6.2) correction jobs, respectively. Initial particle picking was performed using the Gaussian Blob Picker (from cryoSPARC 4.6.2), resulting in a total of 1,949,389 particle picks. After several rounds of iterative 2D classification, 16,120 particles were used to generate an ab initio model without enforcing symmetry (C1).

Selected 2D classes were subsequently used as templates for reference-based particle picking, yielding 2,091,825 particle locations. Particles were extracted with a 400 px box and fourfold binning applied, resulting in a final pixel size of 3.32 Å/px. After iterative 2D classification, 355,436 selected particles were re-extracted with a 400 px box and no binning applied. The re-extracted particles and the ab initio model generated previously were used to perform a round of non-uniform refinement with C2 symmetry enforced, before subsequently subjecting the aligned particles to a 3D classification job with 6 classes, resulting in 4 practically empty junk classes ($\leq$32 particles) and 2 classes resembling the h20S proteasome. The best 3D class, comprised of 209,994 particles, was selected and subjected to homogenous refinement with C2 symmetry enforced, followed by reference-based motion correction. We rejected 257 particles due to their proximity to the micrograph edge, and the remaining 209,737 particles were subjected to global and local CTF refinement before performing a final homogenous refinement with C2 symmetry, resulting in the final reconstruction with an $FSC_{0.143}$ of 2.55 Å (Table S1). Finally, the handedness of the resulting map was flipped using the Volume Tool utility within cryoSPARC before sharpening the map using the EMReady software (version 1.2) [22]. An overview of the cryo-EM data processing workflow is shown in Figure S1.

4.4. Modelling and Refinement

A previously determined cryo-EM structure of the human 20S proteasome (PDB: 7PG9) [37] was used as a starting model and initially docked into the sharpened density map using ChimeraX (version 1.7.1) [38]. Next, the model was iteratively refined using a combination of automated real-space refinement using Phenix real_space_refine [39], manual refinement in Coot 0.9.8.95 [40,41] and structure optimization using the ISOLDE [42] package in ChimeraX version 1.7.1 until satisfactory validation metric and map-correlation had been achieved. Refinement restraints for MZB were generated using Jligand [43]. Model validation was performed with MolProbity [44]. Visualisation was performed in ChimeraX version 1.7.1 and PyMOL version 3.0.1 (Schrödinger, LLC, New York, NY, USA).

Supplementary Materials: The following supporting information can be downloaded at: https://www.mdpi.com/article/10.3390/molecules30061386/s1, Figure S1. Workflow of cryo-EM data processing; Figure S2. Cryo-EM single particle analysis of h20S in complex with MZB; Figure S3. Detailed views of the cryo-EM maps in active sites of human 20S with covalently bound MZB inhibitor; Figure S4. 2D representation of the interactions between small molecule inhibitor MZB and the h20S active sites; Figure S5. Half-maximal inhibitory concentration (IC_{50}) of MZB for the proteolytic *Tv*20S subunits; Table S1: Cryo-EM data collection, refinement and validation statistics [22,36,45].

Author Contributions: J.S. and E.B. conceived the project. J.S. and H.S. collected and processed cryo-EM data, P.F. performed enzyme experiments. H.S., P.F., A.J.O., J.S. and E.B. wrote, revised and edited the manuscript. P.F., A.J.O., J.S. and E.B obtained funding. All authors have read and agreed to the published version of the manuscript.

Funding: This research was funded by the project the National Institute Virology and Bacteriology (Programme EXCELES, Project No. LX22NPO5103)—funded by the European Union—Next Generation EU. We also acknowledge the CF BIC of CIISB, Instruct-CZ Centre, supported by MEYS CR (LM2023042) and European Regional Development Fund-Project "Innovation of Czech Infrastructure for Integrative Structural Biology" (No. CZ.02.01.01/00/23_015/0008175). P.F. received funding from the European Union's Horizon 2020 research and innovation programme under the Marie Skłodowska-Curie grant, agreement No. [846688]. This research was also supported by NIH awards R01AI158612, R21AI146387, R21AI133393 and R21AI171824 to A.J.O. We thank Dr Jehad Almaliti for his explanation of the mechanism of action of MZB.

Institutional Review Board Statement: Not applicable.

Informed Consent Statement: Not applicable.

Data Availability Statement: The atomic model of the presented complex of human 20S proteasome with small molecule inhibitor MZB has been deposited in the Protein Data Bank under accession code 9HMN. The associated cryo-EM density maps, half-maps and masks have been deposited in the Electron Microscopy Data Bank under accession code EMD-52296. The starting model used for the modelling of the complex is deposited in the Protein Data Bank under accession code 7PG9.

Acknowledgments: We thank and express our gratitude to Tomas Kouba and Anatolij Filimonenko for their assistance in the cryo-EM data collection.

Conflicts of Interest: The authors declare no competing interests.

References

1. Feling, R.H.; Buchanan, G.O.; Mincer, T.J.; Kauffman, C.A.; Jensen, P.R.; Fenical, W. Salinosporamide A: A highly cytotoxic proteasome inhibitor from a novel microbial source, a marine bacterium of the new genus *Salinospora*. *Angew. Chem. Int. Ed.* **2003**, *42*, 355–357. [CrossRef]
2. Mincer, T.J.; Jensen, P.R.; Kauffman, C.A.; Fenical, W. Widespread and Persistent Populations of a Major New Marine Actinomycete Taxon in Ocean Sediments. *Appl. Environ. Microbiol.* **2002**, *68*, 5005–5011. [CrossRef] [PubMed]

3. Roth, P.; Gorlia, T.; Reijneveld, J.C.; de Vos, F.; Idbaih, A.; Frenel, J.-S.; Le Rhun, E.; Sepulveda, J.M.; Perry, J.; Masucci, G.L.; et al. Marizomib for patients with newly diagnosed glioblastoma: A randomized phase 3 trial. *Neuro-Oncology* **2024**, *26*, 1670–1682. [CrossRef]
4. Di, K.; Lloyd, G.K.; Abraham, V.; MacLaren, A.; Burrows, F.J.; Desjardins, A.; Trikha, M.; Bota, D.A. Marizomib activity as a single agent in malignant gliomas: Ability to cross the blood-brain barrier. *Neuro-Oncology* **2015**, *18*, 840–848. [CrossRef]
5. Manton, C.A.; Johnson, B.; Singh, M.; Bailey, C.P.; Bouchier-Hayes, L.; Chandra, J. Induction of cell death by the novel proteasome inhibitor marizomib in glioblastoma in vitro and in vivo. *Sci. Rep.* **2016**, *6*, 18953. [CrossRef]
6. Manasanch, E.E.; Korde, N.; Zingone, A.; Tageja, N.; de Larrea, C.F.; Bhutani, M.; Wu, P.; Roschewski, M.; Landgren, O. The proteasome: Mechanisms of biology and markers of activity and response to treatment in multiple myeloma. *Leuk. Lymphoma* **2014**, *55*, 1707–1714. [CrossRef]
7. Eisenberg-Lerner, A.; Benyair, R.; Hizkiahou, N.; Nudel, N.; Maor, R.; Kramer, M.P.; Shmueli, M.D.; Zigdon, I.; Lev, M.C.; Ulman, A.; et al. Golgi organization is regulated by proteasomal degradation. *Nat. Commun.* **2020**, *11*, 409. [CrossRef]
8. Tanaka, K. The proteasome: Overview of structure and functions. *Proc. Jpn. Acad Ser. B Phys. Biol. Sci.* **2009**, *85*, 12–36. [CrossRef]
9. Coux, O.; Tanaka, K.; Goldberg, A.L. STRUCTURE AND FUNCTIONS OF THE 20S AND 26S PROTEASOMES. *Annu. Rev. Biochem.* **1996**, *65*, 801–847. [CrossRef]
10. Lupas, A.; Zwickl, P.; Wenzel, T.; Seemuller, E.; Baumeister, W. Structure and Function of the 20S Proteasome and of Its Regulatory Complexes. *Cold Spring Harb. Symp. Quant. Biol.* **1995**, *60*, 515–524. [CrossRef]
11. Löwe, J.; Stock, D.; Jap, B.; Zwickl, P.; Baumeister, W.; Huber, R. Crystal Structure of the 20S Proteasome from the Archaeon *T. acidophilum* at 3.4 Å Resolution. *Science* **1995**, *268*, 533–539. [CrossRef] [PubMed]
12. Orlowski, M.; Cardozo, C.; Michaud, C. Evidence for the presence of five distinct proteolytic components in the pituitary multicatalytic proteinase complex. Properties of two components cleaving bonds on the carboxyl side of branched chain and small neutral amino acids. *Biochemistry* **1993**, *32*, 1563–1572. [CrossRef]
13. Mishra, R.; Upadhyay, A.; Prajapati, V.K.; Mishra, A. Proteasome-mediated proteostasis: Novel medicinal and pharmacological strategies for diseases. *Med. Res. Rev.* **2018**, *38*, 1916–1973. [CrossRef]
14. Schmidt, M.; Finley, D. Regulation of proteasome activity in health and disease. *Biochim. Biophys. Acta (BBA)-Mol. Cell Res.* **2014**, *1843*, 13–25. [CrossRef]
15. Fenteany, G.; Standaert, R.F.; Lane, W.S.; Choi, S.; Corey, E.J.; Schreiber, S.L. Inhibition of Proteasome Activities and Subunit-Specific Amino-Terminal Threonine Modification by Lactacystin. *Science* **1995**, *268*, 726–731. [CrossRef]
16. Groll, M.; Potts, B.C. Proteasome Structure, Function, and Lessons Learned from Beta-Lactone Inhibitors. *Curr. Top. Med. Chem.* **2011**, *11*, 2850–2878. [CrossRef]
17. Macherla, V.R.; Mitchell, S.S.; Manam, R.R.; Reed, K.A.; Chao, T.-H.; Nicholson, B.; Deyanat-Yazdi, G.; Mai, B.; Jensen, P.R.; Fenical, W.F.; et al. Structure–Activity Relationship Studies of Salinosporamide A (NPI-0052), a Novel Marine Derived Proteasome Inhibitor. *J. Med. Chem.* **2005**, *48*, 3684–3687. [CrossRef]
18. Groll, M.; Huber, R.; Potts, B.C.M. Crystal Structures of Salinosporamide A (NPI-0052) and B (NPI-0047) in Complex with the 20S Proteasome Reveal Important Consequences of β-Lactone Ring Opening and a Mechanism for Irreversible Binding. *J. Am. Chem. Soc.* **2006**, *128*, 5136–5141. [CrossRef]
19. Silhan, J.; Silhan, J.; Fajtova, P.; Fajtova, P.; Bartosova, J.; Bartosova, J.; Hurysz, B.M.; Hurysz, B.M.; Almaliti, J.; Almaliti, J.; et al. Structural elucidation of recombinant *Trichomonas vaginalis* 20S proteasome bound to covalent inhibitors. *Nat. Commun.* **2024**, *15*, 8621. [CrossRef]
20. Prudhomme, J.; McDaniel, E.; Ponts, N.; Bertani, S.; Fenical, W.; Jensen, P.; Le Roch, K. Marine Actinomycetes: A New Source of Compounds against the Human Malaria Parasite. *PLoS ONE* **2008**, *3*, e2335. [CrossRef]
21. Afonine, P.V.; Klaholz, B.P.; Moriarty, N.W.; Poon, B.K.; Sobolev, O.V.; Terwilliger, T.C.; Adams, P.D.; Urzhumtsev, A. New tools for the analysis and validation of cryo-EM maps and atomic models. *Acta Crystallogr. Sect. D Struct. Biol.* **2018**, *74*, 814–840. [CrossRef]
22. He, J.; Li, T.; Huang, S.-Y. Improvement of cryo-EM maps by simultaneous local and non-local deep learning. *Nat. Commun.* **2023**, *14*, 3217. [CrossRef]
23. Borissenko, L.; Groll, M. 20S Proteasome and Its Inhibitors: Crystallographic Knowledge for Drug Development. *Chem. Rev.* **2007**, *107*, 687–717. [CrossRef] [PubMed]
24. Fajtova, P.; Hurysz, B.M.; Miyamoto, Y.; Serafim, M.S.M.; Jiang, Z.; Vazquez, J.M.; Trujillo, D.F.; Liu, L.J.; Somani, U.; Almaliti, J.; et al. Distinct substrate specificities of the three catalytic subunits of the *Trichomonas vaginalis* proteasome. *Protein Sci.* **2024**, *33*, e5225. [CrossRef]
25. Ciechanover, A. The ubiquitin-proteasome proteolytic pathway. *Cell* **1994**, *79*, 13–21. [CrossRef]

26. Chauhan, D.; Catley, L.; Li, G.; Podar, K.; Hideshima, T.; Velankar, M.; Mitsiades, C.; Mitsiades, N.; Yasui, H.; Letai, A.; et al. A novel orally active proteasome inhibitor induces apoptosis in multiple myeloma cells with mechanisms distinct from Bortezomib. *Cancer Cell* **2005**, *8*, 407–419. [CrossRef]
27. Nunes, A.T.; Annunziata, C.M. Proteasome inhibitors: Structure and function. *Semin. Oncol.* **2017**, *44*, 377–380. [CrossRef]
28. Park, J.E.; Miller, Z.; Jun, Y.; Lee, W.; Kim, K.B. Next-generation proteasome inhibitors for cancer therapy. *Transl. Res.* **2018**, *198*, 1–16. [CrossRef]
29. Kegyes, D.; Gulei, D.; Drula, R.; Cenariu, D.; Tigu, B.; Dima, D.; Tanase, A.; Badelita, S.; Buzoianu, A.-D.; Ciurea, S.; et al. Proteasome inhibition in combination with immunotherapies: State-of-the-Art in multiple myeloma. *Blood Rev.* **2023**, *61*, 101100. [CrossRef]
30. Raninga, P.V.; Lee, A.; Sinha, D.; Dong, L.-F.; Datta, K.K.; Lu, X.; Croft, P.K.-D.; Dutt, M.; Hill, M.; Pouliot, N.; et al. Marizomib suppresses triple-negative breast cancer via proteasome and oxidative phosphorylation inhibition. *Theranostics* **2020**, *10*, 5259–5275. [CrossRef]
31. Zhang, Z.; Zhang, S.; Lin, B.; Wang, Q.; Nie, X.; Shi, Y. Combined treatment of marizomib and cisplatin modulates cervical cancer growth and invasion and enhances antitumor potential in vitro and in vivo. *Front. Oncol.* **2022**, *12*, 974573. [CrossRef]
32. Kaplan, G.S.; Torcun, C.C.; Grune, T.; Ozer, N.K.; Karademir, B. Proteasome inhibitors in cancer therapy: Treatment regimen and peripheral neuropathy as a side effect. *Free Radic. Biol. Med.* **2017**, *103*, 1–13. [CrossRef] [PubMed]
33. Robbertse, L.; Fajtová, P.; Šnebergerová, P.; Jalovecká, M.; Levytska, V.; da Silva, E.B.; Sharma, V.; Pachl, P.; Almaliti, J.; Al-Hindy, M.; et al. Evaluating Antimalarial Proteasome Inhibitors for Efficacy in *Babesia* Blood Stage Cultures. *ACS Omega* **2024**, *9*, 44989–44999. [CrossRef] [PubMed]
34. Eadsforth, T.C.; Torrie, L.S.; Rowland, P.; Edgar, E.V.; MacLean, L.M.; Paterson, C.; Robinson, D.A.; Shepherd, S.M.; Thomas, J.; Thomas, M.G.; et al. Pharmacological and structural understanding of the Trypanosoma cruzi proteasome provides key insights for developing site-specific inhibitors. *J. Biol. Chem.* **2024**, *301*, 108049. [CrossRef]
35. Miller, C.P.; Manton, C.A.; Hale, R.; DeBose, L.; Macherla, V.R.; Potts, B.C.; Palladino, M.A.; Chandra, J. Specific and prolonged proteasome inhibition dictates apoptosis induction by marizomib and its analogs. *Chem. Interact.* **2011**, *194*, 58–68. [CrossRef]
36. Punjani, A.; Rubinstein, J.L.; Fleet, D.J.; A Brubaker, M. cryoSPARC: Algorithms for rapid unsupervised cryo-EM structure determination. *Nat. Methods* **2017**, *14*, 290–296. [CrossRef]
37. Sahu, I.; Mali, S.M.; Sulkshane, P.; Xu, C.; Rozenberg, A.; Morag, R.; Sahoo, M.P.; Singh, S.K.; Ding, Z.; Wang, Y.; et al. The 20S as a stand-alone proteasome in cells can degrade the ubiquitin tag. *Nat. Commun.* **2021**, *12*, 6173. [CrossRef]
38. Pettersen, E.F.; Goddard, T.D.; Huang, C.C.; Meng, E.C.; Couch, G.S.; Croll, T.I.; Morris, J.H.; Ferrin, T.E. UCSF ChimeraX: Structure visualization for researchers, educators, and developers. *Protein Sci.* **2020**, *30*, 70–82. [CrossRef]
39. Afonine, P.V.; Poon, B.K.; Read, R.J.; Sobolev, O.V.; Terwilliger, T.C.; Urzhumtsev, A.; Adams, P.D. Real-space refinement in *PHENIX* for cryo-EM and crystallography. *Acta Crystallogr. Sect. D Struct. Biol.* **2018**, *74*, 531–544. [CrossRef]
40. Liebschner, D.; Afonine, P.V.; Baker, M.L.; Bunkóczi, G.; Chen, V.B.; Croll, T.I.; Hintze, B.; Hung, L.-W.; Jain, S.; McCoy, A.J.; et al. Macromolecular structure determination using X-rays, neutrons and electrons: Recent developments in *Phenix*. *Acta Crystallogr. Sect. D Struct. Biol.* **2019**, *75*, 861–877. [CrossRef]
41. Emsley, P.; Lohkamp, B.; Scott, W.G.; Cowtan, K. Features and development of Coot. *Acta Crystallogr. Sect. D Biol. Crystallogr.* **2010**, *66*, 486–501. [CrossRef] [PubMed]
42. Croll, T.I. *ISOLDE*: A physically realistic environment for model building into low-resolution electron-density maps. *Acta Crystallogr. Sect. D Struct. Biol.* **2018**, *74*, 519–530. [CrossRef] [PubMed]
43. Lebedev, A.A.; Young, P.; Isupov, M.N.; Moroz, O.V.; Vagin, A.A.; Murshudov, G.N. *JLigand*: A graphical tool for the *CCP4* template-restraint library. *Acta Crystallogr. Sect. D Struct. Biol.* **2012**, *68*, 431–440. [CrossRef] [PubMed]
44. Williams, C.J.; Headd, J.J.; Moriarty, N.W.; Prisant, M.G.; Videau, L.L.; Deis, L.N.; Verma, V.; Keedy, D.A.; Hintze, B.J.; Chen, V.B.; et al. MolProbity: More and better reference data for improved all-atom structure validation. *Protein Sci.* **2018**, *27*, 293–315. [CrossRef]
45. Laskowski, R.A.; Swindells, M.B. LigPlot+: Multiple ligand–protein interaction diagrams for drug discovery. *J. Chem. Inf. Model.* **2011**, *51*, 2778–2786. [CrossRef]

molecules

MDPI

Article

Integrating Surface Plasmon Resonance and Docking Analysis for Mechanistic Insights of Tryptase Inhibitors

Alessia Porta [1], Candida Manelfi [2], Carmine Talarico [2], Andrea Rosario Beccari [2], Margherita Brindisi [3], Vincenzo Summa [3], Daniela Iaconis [2,*], Marco Gobbi [1,*] and Marten Beeg [1]

[1] Istituto di Ricerche Farmacologiche Mario Negri IRCCS, Via Mario Negri, 2, 20156 Milano, Italy; alessia.porta@marionegri.it (A.P.); marten.beeg@marionegri.it (M.B.)

[2] EXSCALATE-Dompé Farmaceutici SpA, via Tommaso De Amicis 95, 80131 Napoli, Italy; candida.manelfi@dompe.com (C.M.); carmine.talarico@dompe.com (C.T.); andrea.beccari@dompe.com (A.R.B.)

[3] Department of Pharmacy, School of Medicine and Surgery, University of Naples Federico II, Via D. Montesano 49, 80131 Naples, Italy; margherita.brindisi@unina.it (M.B.); vincenzo.summa@unina.it (V.S.)

* Correspondence: daniela.iaconis@dompe.com (D.I.); marco.gobbi@marionegri.it (M.G.)

Abstract: Tryptase is a tetrameric serine protease and a key component of mast cell granules. Here, we explored an integrated approach to characterize tryptase ligands, combining novel experimental binding studies using Surface Plasmon Resonance, with in silico analysis through the Exscalate platform. For this, we focused on three inhibitors previously reported in the literature, including a bivalent inhibitor and its corresponding monovalent compound. All three ligands showed concentration-dependent binding to immobilized human tryptase with the bivalent inhibitor showing the highest affinity. Furthermore, Rmax values were similar, indicating that the compounds occupy all four binding pockets of the tryptase tetramer. This hypothesis was supported by in silico computational analysis that revealed the binding mode of the monovalent ligand, one in each monomer pocket, compared with crystal structure of the bivalent one, which simultaneously occupies two binding pockets. Additionally, we solved the 2.06 Å X-ray crystal structures of human Tryptase beta-2 (hTPSB2), in both its apo form and in complex with compound #**1**, experimentally confirming the binding mode and the key molecular interactions predicted by docking studies for this compound. This integrated approach offers a robust framework for elucidating both the strength and mode of interaction of potential tryptase inhibitors.

Keywords: tryptase inhibitors; molecular interactions; molecular modeling; surface plasmon resonance; in silico docking; X-ray crystal structures

1. Introduction

Mast cells are immune cells derived from myeloid progenitors that play a crucial role in inflammatory responses, neuroimmune regulation, and tissue remodeling. These cells contain granules rich in bioactive molecules, including histamine, heparin, and proteases, which are released upon activation [1]. Among these proteases, tryptase, a tetrameric serine protease, is one of the most abundant and functionally significant enzymes in mast cell granules [2,3]. Tryptase is implicated in extracellular matrix (ECM) remodeling and inflammatory diseases, including asthma, chronic pain, fibrosis, and cancer progression [4,5]. Due to its central role in mast cell-mediated responses, tryptase inhibition has emerged as a promising therapeutic strategy for conditions linked to excessive mast cell activation [6,7].

Despite the strong therapeutic interest in tryptase inhibitors, a detailed understanding of their binding stoichiometry and mechanism of action remains incomplete. While many inhibitors have been evaluated based on potency, their exact mode of interaction with the tryptase tetramer is not well characterized at the molecular level. In particular, bivalent inhibitors—which engage multiple active sites simultaneously—have shown increased potency compared to monovalent compounds, yet the molecular basis of this enhanced affinity is not fully understood.

To address this gap, we employed an integrated approach that combines computational and experimental methodologies to investigate both binding affinity and stoichiometry of tryptase inhibitors, with the additional aim to cross-validate these methodologies for more reliable screening of novel compounds. For this, in the present study we resynthesized and examined three tryptase inhibitors reported in the literature (Figure 1) [8,9].

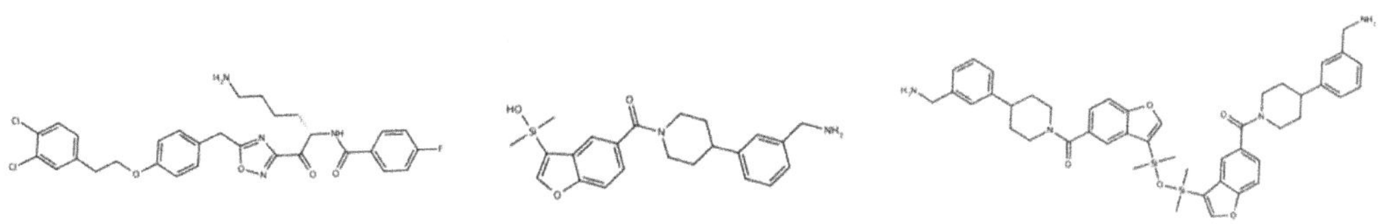

compound #1	compound #2m	compound #2d
599.48 Da	408.57 Da	799.13 Da
NCCCC[C@H](NC(=O)c1ccc(F)cc1)C(=O)c2noc(Cc3ccc(OCCc4ccc(Cl)c(Cl)c4)cc3)n2	C[Si](C)(O)c1coc2ccc(cc12)C(=O)N3CCC(CC3)c4cccc(CN)c4.OC=O	C[Si](C)(O[Si](C)(C)c1coc2ccc(cc12)C(=O)N3CCC(CC3)c4cccc(CN)c4)c5coc6ccc(cc56)C(=O)N7CCC(CC7)c8cccc(CN)c8.OC=O.OC=O
(a)	(b)	(c)

Figure 1. Structure, molecular weight and SMILE of compounds **#1** (**a**), **#2m** (**b**), and **#2d** (**c**).

To experimentally determine the binding properties, we exploited Surface Plasmon Resonance (SPR) technology, which allows us to measure the association and dissociation rate constants and, as a consequence, to provide information on the binding mode (e.g., on the presence of avidity effects, possibly exhibited by bivalent inhibitors) and on the binding stoichiometry [10,11]. It was already described that bivalent inhibitors usually enhance binding affinity compared to the monovalent ones by exploiting the ability to occupy two binding sites on a target molecule simultaneously through a cooperative effect [12]. Gathering this information during drug discovery helps to design more effective drugs, and the SPR approach provides these data quickly and cost effectively. Molecular docking simulations with the Exscalate platform were then applied to obtain the molecular details of the interaction of these inhibitors within the tryptase tetramer and to support the experimental findings. In order to further validate our findings, the X-ray crystal structure of one inhibitor–tryptase complex was solved, confirming the data.

By integrating in silico predictions with biophysical and structural validation, this study offers new insights into the molecular mechanisms of tryptase inhibition. The findings not only enhance our understanding of ligand–tryptase interactions but also support the rational design of more potent and selective tryptase inhibitors for therapeutic applications.

2. Results

2.1. Surface Plasmon Resonance Experimental Validation

For the present studies, aiming at investigating the binding properties of small molecules flowing over human tryptase immobilized on the sensor chip, we used a highly sensitive SPR instrument, and a chip allowing high immobilization levels. The experimental conditions were optimized, as described in the Methods section, to have adequate and reproducible binding signals.

Each of the three investigated compounds (Figure 1: **#1**, **#2m**, **#2d**) was tested in at least three independent SPR experimental sessions and injected at different concentrations.

Representative sensorgrams are shown in Figure 2a–c to highlight the concentration-dependent binding and the good interpolation of the data with a global fitting (i.e., considering both the association and the dissociation phases at all the compound's concentrations). For these fitting, we needed to use an equation including two binding sites, suggesting the presence of a specific high-affinity interaction occurring at lower drug concentrations, while at higher concentrations, a lower affinity interaction also appears. Figure 2d reports representative sensorgrams obtained with a saturating concentration of the compounds for a more reliable determination of the Rmax values.

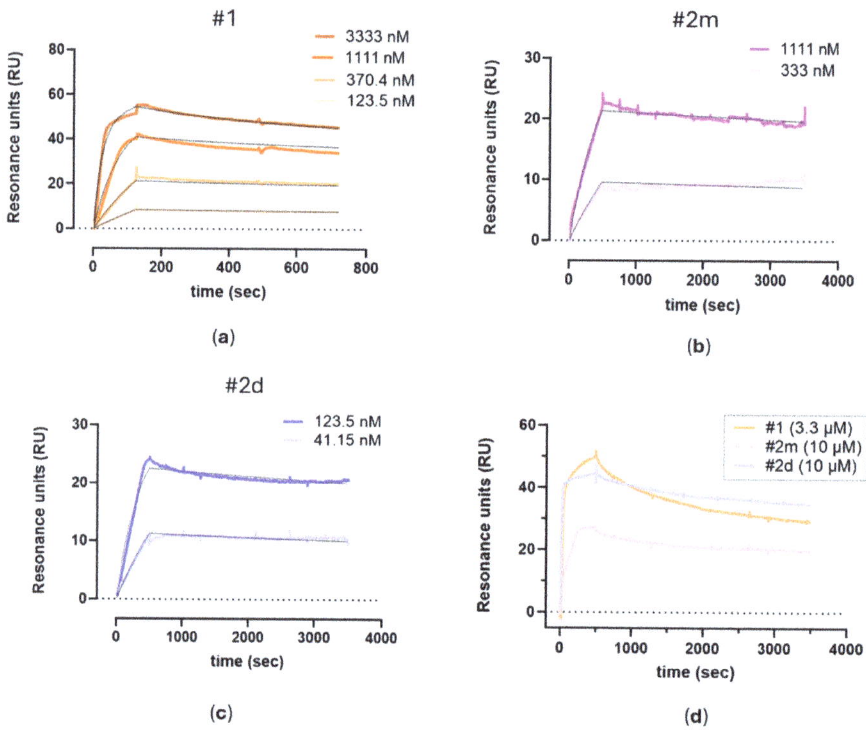

Figure 2. Representative SPR sensorgrams (i.e., the time course of SPR binding signals, in Resonance Units, RU) were obtained by injecting different concentrations of **#1** (**a**), **#2m** (**b**), and **#2d** (**c**) over human tryptase immobilized on the sensor chip. For each compound, the sensorgrams were fitted globally using a two-site binding model, with the resulting fitting curves represented by black lines. (**d**) Overlay of SPR sensorgrams at saturating concentrations: 3.3 μM for compound **#1** and 10 μM for compounds **#2m** and **#2d**.

The affinity constants of the high-affinity component obtained in each independent experiment, together with the mean values, are reported in Table 1 to allow for appreciation of the inter-assay variability. The high-affinity site was chosen as it accounts for the majority of interactions and is therefore possibly more relevant to the overall binding mechanism. The compound with the highest affinity (K_D = 2.1 nM) was the bivalent inhibitor **#2d**, whereas the one with the lowest affinity (K_D = 81 nM) was the corresponding monovalent compound **#2m**, while compound **#1** showed an intermediate affinity (K_D = 12.3 nM). Considering the two monomers, the higher affinity of compound **#1** in comparison with compound **#2m** is mainly due to a 17-fold faster association, partially counteracted by a 3-fold faster dissociation; the comparison of monomeric compound **#2m** with the corresponding dimeric compound **#2d** shows that the higher affinity of the latter is also mainly due to a faster association (38-fold) while dissociation constants are similar.

Table 1. Summary of the kinetic constants obtain from the fitting curves in different experimental sessions. The last row for each compound represents the mean ± standard deviation (SD) of the individual measurements.

Compound	k_a [1/Ms]	k_d [1/s]	K_D [nM]	Rmax [RU]
#1	1.187×10^4	2.6×10^{-4}	22.0	48.0
	1.287×10^4	0.8×10^{-4}	6.5	45.8
	1.143×10^4	1.0×10^{-4}	8.5	38.6
	$1.2 \pm 0.08 \times 10^4$	$1.5 \pm 0.99 \times 10^{-4}$	12.3 ± 8.4	44.1 ± 4.9
#2m	729.0	7.6×10^{-5}	105.0	29.3
	721.0	2.1×10^{-5}	29.1	22.4
	633.1	7.2×10^{-5}	114.0	27.2
	694 ± 53.2	$5.3 \pm 3 \times 10^{-5}$	81 ± 46.4	26.3 ± 3.5
#2d	2.611×10^4	8.5×10^{-5}	3.2	27.4
	2.638×10^4	5.0×10^{-5}	1.9	40.0
	2.931×10^4	4.7×10^{-5}	1.7	34.1
	2.397×10^4	4.0×10^{-5}	1.7	29.6
	$2.7 \pm 0.21 \times 10^4$	$5.6 \pm 2.01 \times 10^{-5}$	2.1 ± 0.7	32.8 ± 5.6

As reported in Table 1, the Rmax value of the monomer **#2m** (26.3 ± 3.5 RU, MW409) was similar to the Rmax value of the dimer **#2d** (32.8 ± 5.6 RU, MW 800). A similar Rmax can be also calculated for compound **#1**, taking into account that its MW (600 RU) is 1.5-fold that of compound **#2m**, resulting in a normalized Rmax of 29 ± 3.2 RU. These data indicate that all four binding pockets of the tryptase tetramer are occupied but with different stoichiometric ratios. Monovalent compounds (**#1** and **#2m**) follow a 4:1 stoichiometry, with four molecules binding per enzyme, whereas the bivalent compound (**#2d**) follows a 2:1 stoichiometry, binding two molecules per enzyme. This suggests that monovalent inhibitors engage each active site independently, while the bivalent inhibitor spans two adjacent sites.

2.2. Molecular Docking Supports Affinity and Binding Stoichiometry of Tryptase Inhibitors

The SPR studies provided insights into the binding kinetics, affinities, and binding stoichiometry of compounds **#1**, **#2m**, and **#2d**. To further characterize these interactions, docking studies were performed to explore their binding modes.

While the binding mode of bivalent inhibitor **#2d** has been previously characterized (PDB: 4MPW) [8], the exact interaction patterns of monovalent inhibitors, including compounds **#1** and **#2m**, required further investigation.

In order to explore and validate these binding modes through molecular docking, we generated X-ray crystal structures of human tryptase beta-2 (hTPSB2) in its apo form.

The structure was solved at a resolution of 2.06 Å and displayed the typical trypsin-like serine protease fold. It forms a homotetramer, with the catalytic sites facing toward the central channel (Figure 3).

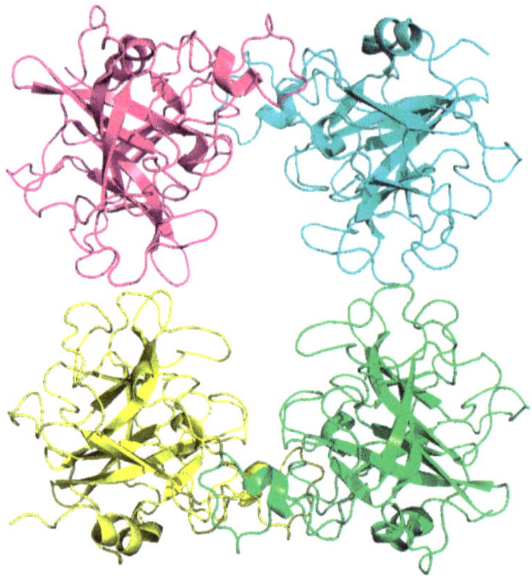

Figure 3. Overall structure of hTPSB2 tetramer shown as ribbon diagram.

The docking studies presented in Figure 4 revealed distinct binding interactions among the inhibitors, with compound **#1** exhibiting a higher number of polar interactions, five hydrogen bonds with key catalytic site residues instead of the three ones estabilished by compound **#2m**, and a more extended network of hydrophobic contacts, reaching a new larger portion of the protein, compared to compound **#2m**. Also, comparison between compound **#2m** and compound **#2d** shows a more efficient binding mode with additional hydrophobic interactions involving the central disiloxane linker of the bivalent compound **#2d**, in agreement with experimental results showing an increase in the binding affinity.

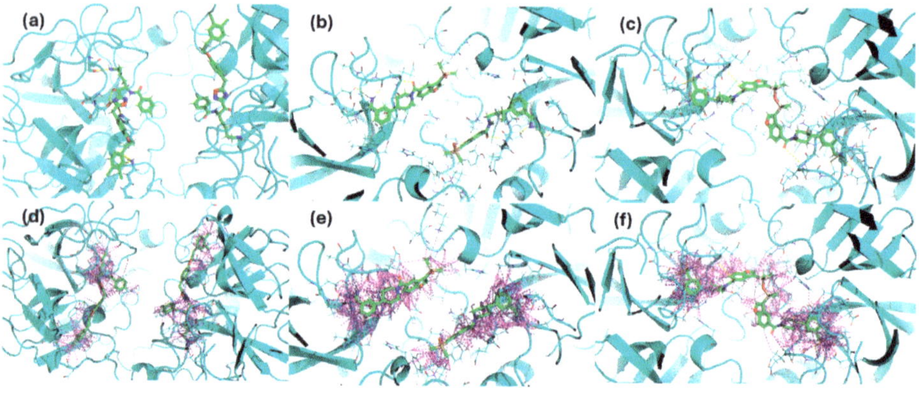

Figure 4. Docking-predicted binding modes of compound **#1** (**a**,**d**), compound **#2m** (**b**,**e**), and compound **#2d** (**c**,**f**), within the tryptase active site respectively showing H-bond networks (yellow dashed lines panel (**a**–**c**)) and hydrophobic contacts (purple dashed lines panel (**d**–**f**)).

In-depth analysis of compound **#1** revealed that it engages key catalytic site residues via hydrogen bonds to the side chain atoms of Ser209, His59, and Lys66, as well as hydrogen bonds to the main chain atoms of Gly207 and Ser204 (Figure 5).

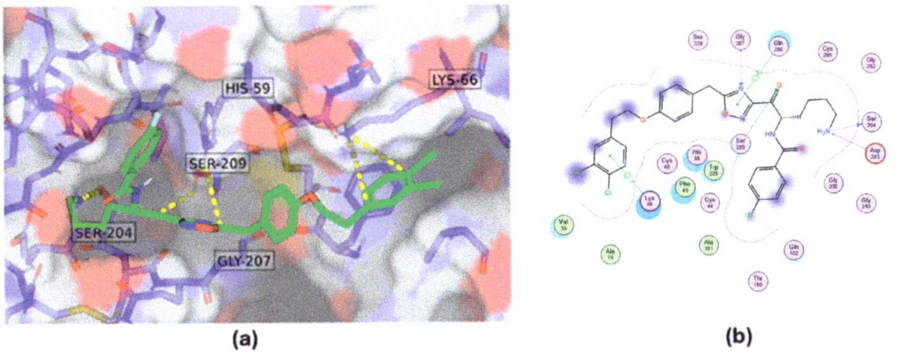

Figure 5. Docking binding mode showing interactions of compound **#1** with key residues of tryptase. Three-dimensional binding interaction representation (**a**) and two-dimensional binding interaction map (**b**).

In particular, the H-bond network between the ketone group of the ligand and the hydroxyl group of the residue Ser209 suggests a potential covalent binding mode for compound **#1** [13]. In order to verify this hypothesis, we applied a covalent docking procedure confirming our assumption: the covalent mechanism is favored and the binding mode is very similar to the previous one described.

To experimentally confirm the effective mechanism of action of compound **#1**, we have solved the crystal structure of its complex with tryptase (Figure 6).

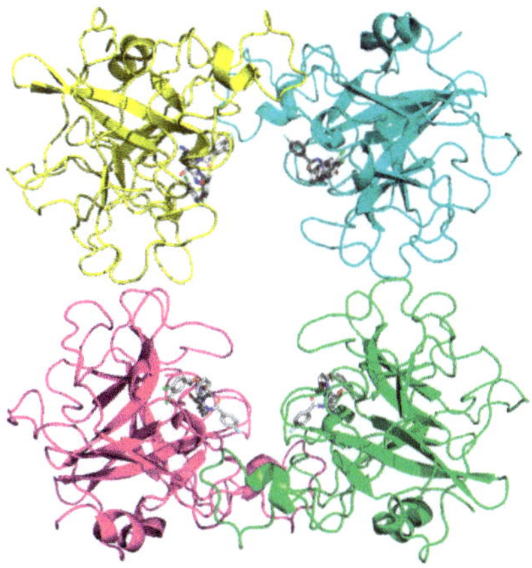

Figure 6. Overall structure of hTPSB2, containing compound **#1**, shown as ribbon diagram.

Based on a distance of less than 3.5 Å of the donor and acceptor atoms, we identify nine specific hydrogen bonds of the ligand compound **#1**, namely to the main chain atoms of Ser204, Ser228, Cys205, Gly207, and Ser209, as well as the side chain atoms of Ser204, Gln206, and His59. According to the above distance criteria, we also suggest the presence of additional hydrophilic interactions with the main chain atoms of Trp229, Ile241, Gly232, and Arg238. The crystal structure of the complex confirms the binding mode hypothesized by docking studies, retrieving almost all the key interactions (Figure 7).

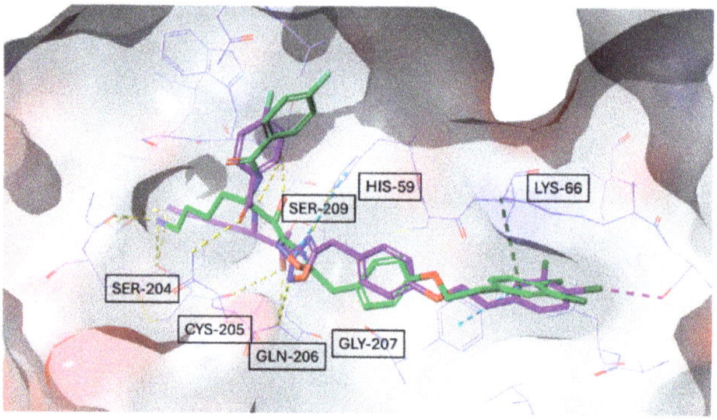

Figure 7. Overlay of the docking-predicted binding mode of compound **#1** (green) with its experimentally determined crystal structure (violet).

The experimental electron density clearly shows that the compound **#1** is covalently bound with C12 of the electrofilic carbonyl group with the hydroxyl group of the active site serine residue, Ser209, forming a reversible hemiketal (hemiacetal), in agreement with the SPR-derived Kd values. In particular, it represents an example of covalent-reversible inhibition of serine protease (Figure 8).

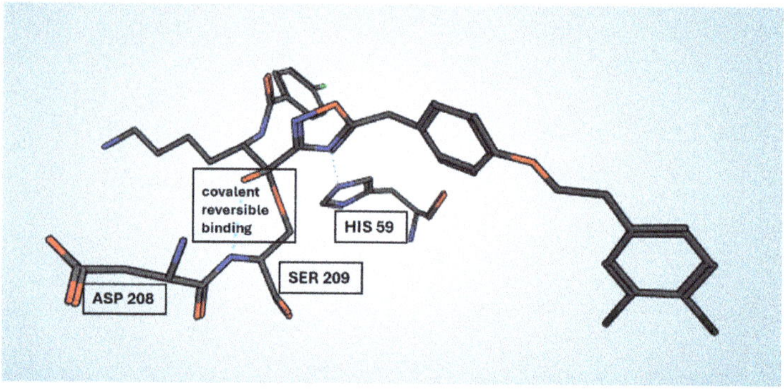

Figure 8. Focus on the reversible covalent binding mode of compound **#1** (violet) in the X-ray complex structure with tryptase.

3. Discussion

The integrated SPR and computational approaches described here offer a robust framework for elucidating both the strength and mode of interaction of potential tryptase inhibitors.

To our knowledge, this is the first study determining the binding constants of tryptase inhibitors, whose activity is usually measured by functional assays. For this aim, we developed a novel SPR-based assay in which the inhibitors flow over human tryptase immobilized on a sensor chip. The assay proved to be sensitive and reproducible, with a relatively low inter-assay variability.

In particular, SPR data provided clear evidence for binding stoichiometry—distinguishing monovalent from bivalent binding modes—even in the absence of crystal structures. This capability is crucial in guiding rational drug design, as seen by the dramatic increase in binding affinity when the monomeric ligand (**#2m**) is converted into its dimeric counterpart (**#2d**), yielding a 40-fold improvement. These results correlate well with functional assays (IC_{50} = 2.51 nM for **#2d** vs. 446 nM for **#2m**) [8,9] and illustrate how detailed kinetic and stoichiometric insights can inform more targeted lead optimization.

The binding stoichiometry findings were further supported by structural and computational data. The X-ray crystal structure of compound **#1** confirmed a 4:1 stoichiometry, showing that four molecules bind per tryptase tetramer, each occupying an independent active site (Figure 9a). Similarly, the previously resolved complex of the bivalent inhibitor (PDB: 4MPW) validated a 2:1 stoichiometry, where compound **#2d** spans two adjacent monomers, bridging their active sites (Figure 9c). These findings illustrate that monovalent inhibitors bind independently to each of the four active sites, while bivalent inhibitors engage two adjacent monomers, reducing the overall occupancy ratio.

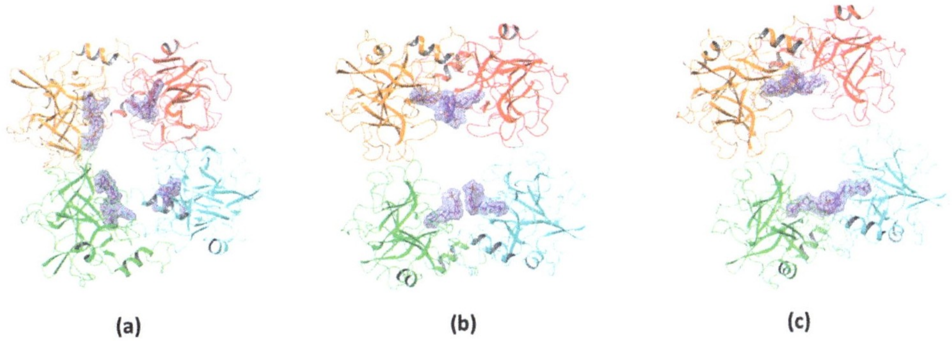

Figure 9. Binding stoichiometry of compounds **#1** (4:1) (**a**), **#2m** (4:1) (**b**), and **#2d** (2:1) (**c**) within the tryptase tetramer.

While direct structural confirmation for compound **#2m** is not yet available, docking results indicate that its binding mode closely resembles that of compound **#1**, supporting the experimentally determined 4:1 stoichiometry (Figure 9b). Additional experimental evidence further supports this hypothesis; in particular, the crystal structure of the tryptase tetramer bound to a monovalent inhibitor (PDB: 2ZEB) shows a binding mode similar to that suggested by docking for compound **#2m** [14]. Together, these findings strongly suggest that compound **#2m** also binds to separate active sites within the tetramer, reinforcing the role of SPR and docking in predicting binding modes even in the absence of crystallographic data.

Furthermore, the docking simulations revealed distinct hydrogen bonding and polar interaction patterns among the inhibitors, providing structural insights into their experimentally observed binding affinities. Compound **#1** exhibited the highest number of hydrogen bonds and polar contacts, forming multiple stabilizing interactions within the tryptase active site. In contrast, compound **#2m** engaged in fewer hydrogen bonds and lacked additional polar interactions, which may contribute to its higher K_D value observed in SPR experiments.

Compound **#2d**, as a bivalent inhibitor, displayed approximately twice the number of hydrogen bonds and polar interactions per molecule compared to compound **#2m**. By spanning two adjacent active sites, it effectively doubled the interaction network, likely resulting in its significantly stronger binding affinity. These findings illustrate how differences in hydrogen bonding and polar contacts correlate with binding strength, reinforcing the role of docking studies in explaining experimentally determined affinities. Building on this, we examined whether bivalent inhibitors induce structural rearrangements in tryptase, as such conformational changes could contribute to the observed kinetic complexity. A comparison of the crystal structures of tryptase complexed with compound **#1** and the previously resolved structure of tryptase bound to compound **#2d** revealed an RMSD value of 0.53 Å, indicating that no significant conformational changes occurred within the tryptase tetramer upon binding either monovalent or bivalent inhibitors. This finding suggests that the multi-step binding behavior observed in SPR arises primarily from the nature of the non-covalent interactions rather than from allosteric effects. However, considering the complex sensorgram fitting required to describe the binding behavior, it remains possible that factors such as heterogeneous immobilization of tryptase or differential exposure of active sites contribute to the observed kinetics. The docking studies not only provided insights into hydrogen bonding and polar interactions but also suggested that compound **#1** could form a reversible covalent bond with the catalytic Ser209 residue. This hypothesis was supported by the proximity of these functional groups in the binding pose, and subsequent covalent docking confirmed a likely reversible covalent interaction mode. This prediction was also confirmed by crystallographic analysis. SPR analysis of the compound **#1** revealed a binding mechanism that could not be fully described by a simple 1:1 interaction model, instead requiring a more complex kinetic fitting approach (two state model). Reversible covalent inhibitors often follow a two-step binding process: an initial non-covalent association, followed by a slower covalent bond formation. This two-step mechanism can generate complex binding kinetics, resembling a two-site binding model even when only a single binding site is involved. Additionally, the data showed a notable dissociation rate, suggesting a dynamic yet stable engagement with the active site. A similar observation was reported by Akçay et al., who studied reversible covalent inhibitors targeting a lysine residue [15]. Their SPR data also required a complex binding model, rather than a simple 1:1 interaction, and showed notable dissociation rates, reinforcing the idea that reversible covalent inhibitors often follow multi-step binding mechanisms. The integration of SPR, docking, and crystallography in this study highlights the power of computational–experimental workflows in drug discovery. While classical docking was used to characterize ligand binding, these approaches could be further enhanced by Artificial Intelligence-Driven Drug Discovery (AIDD) methods. AI and machine learning models, when trained on quantitative kinetic data (SPR), structural insights (X-ray crystallography), and ligand-binding predictions (docking), may have the potential to systematically identify and optimize novel ligands with greater accuracy. Incorporating AI-based strategies could refine virtual screening, improve binding affinity predictions, and streamline structure-based drug design. These advancements represent a natural extension of the computational methods employed in this study, potentially accelerating the identification of improved tryptase inhibitors.

4. Materials and Methods

4.1. Compounds

The protein used for crystallization was recombinant human tryptase beta-2 (hTPSB2), purchased from Promega (Madison, WI, USA; Cat. No. G5631). hTPSB2 was selected for this study based on preliminary screening data indicating its upregulation in a specific tissue of interest, supporting its relevance as a therapeutic target. The inhibitors studied in this work were compound **#1**, compound **#2m**, and compound **#2d**. These compounds were synthesized, purified, and characterized following the literature procedure with slight modifications [8,9].

4.2. Surface Plasmon Resonance Assays

All the SPR studies were carried out by using the very sensitive Sierra SPR-32pro from Bruker (Billerica, MA, USA). Human tryptase B2 (Promega, G5631, Madison, WI, USA) was immobilized via amine coupling on a High-Capacity Amine chip (Bruker, Billerica, MA, USA). After surface activation with EDC (400 mM) and NHS (50 Mm), the protein was injected at 30 μg/mL in acetate buffer pH 5.5 (10 min at 5 μL/min). The residual active sites were then inactivated with 1 M ethanolamine (6 min at 10 μL/min). The final level of the first immobilization was about 10,000 RU.

The three compounds were diluted in PBST 0.1% DMSO at the highest concentration of 10 mM. First, one saturating concentration of each compound (10 μM or 3.3 μM) was injected on the chip surface for 480 s at 25 μL/min. The sensorgrams were baseline corrected by subtracting the signal from an empty reference surface to account for non-specific binding and bulk effect used as reference, and the dissociation constants (Kd) were obtained from the fitting of the sensorgrams with the 1:2 heterogeneous ligands model implemented in the Sierra Analyzer 3.4.5. software. Then, three or more experimental sessions were performed. The three compounds were injected at different concentrations (from 41 nM to 10 μM, at 1:3 serial dilutions) on separate surface spots in parallel. Again, the Sierra Analyzer 3.4.5. software was used to perform a global fitting of the sensorgrams at all the different concentrations to obtain the association and dissociation rate constants (k_a and k_d) and the equilibrium dissociation constant (K_D). All the experimental procedures used in the sessions and the corresponding raw data were saved by the Sierra SPR-32pro instrument and are available.

4.3. Crystallization and Structure Determination of hTPSB2 (Apo and Ligand-Bound Forms)

To investigate the structural basis of tryptase inhibition, we determined X-ray crystal structures of human tryptase beta-2 (hTPSB2) in its apo form and in complex with compound **#1**. These structures were used both as reference models for docking simulations and to provide experimental validation of the predicted binding modes.

Crystallization trials were conducted using a standard screen of approximately 1200 conditions and literature-based refinements. Initial conditions were optimized by systematically varying temperature, protein concentration, drop ratio, pH, and precipitant concentrations to improve crystal quality.

Crystals of apo hTPSB2 and hTPSB2-compound **#1** complex were flash-vitrified and measured at 95 K following PROTEROS Standard Protocols. X-ray diffraction data were collected at the SWISS LIGHT SOURCE (SLS, Villigen, Switzerland) under cryogenic conditions. The crystals belonged to space group P 31, and data were processed using autoPROC, XDS, and AIMLESS (Table 2).

Table 2. Data collection and processing statistics for apo hTPSB2 and hTPSB2-compound **#1** complex.

X-Ray Source	PXII/X10SA (SLS [1])
Wavelength [Å]	1.0001
Detector	Dectris EIGER2 Si 16M
Temperature [K]	100
Space group	P 31
Cell: a; b; c [Å]	78.55; 78.55; 165.69
γ; β; α [°]	90.0; 90.0; 120.0
Resolution [Å]	1.98 (2.02–1.98)
Unique reflections	79,329 (3995)
Multiplicity	6.6 (6.3)
Completeness [%]	100.0 (100.0)
Rpim [%]	7.7 (93.0)
Rsym [%]	18.3 (214.5)
Rmeas [%]	19.9 (234.0)
CC1/2 [%]	99.50 (53.20)
Mean (I)/sd	7.8 (1.3)

[1] SWISS LIGHT SOURCE (SLS, Villigen, Switzerland).

The phase information necessary to determine and analyze the structure was obtained by molecular replacement. A published structure of hTPSB2 was used as a search model. Subsequent model building and refinement was performed according to standard protocols with COOT and the software package CCP4-8, respectively. For the calculation of the free R-factor, a measure to cross-validate the correctness of the final model was used; about 0.5% of measured reflections were excluded from the refinement procedure. Automatically generated local NCS restraints have been applied (keyword "ncsr local" of newer REFMAC5 versions). The water model was built with the "Find waters" algorithm of COOT by putting water molecules in peaks of the Fo-Fc map contoured at 3.0 followed by refinement with REFMAC5 and checking all waters with the validation tool of COOT. The criteria for the list of suspicious waters were B-factor greater 80 Å 2, 2Fo-Fc map less than 1.2 σ, distance to closest contact less than 2.3 Å or more than 3.5 Å. The suspicious water molecules and those in the ligand-binding site (distance to ligand less than 10 Å) were checked manually. The Ramachandran plot of the final model calculated with Molprobity shows 97.21% of all residues in the favored region and 2.79% in the allowed region. There are no outliers in the Ramachandran plot.

The atomic models have been deposited at the Protein Data Bank (PDB) and are available under the accession number 9QFV for the apo structure and 9QFU for the complex with compound **#1**.

4.4. Docking Studies

Molecular docking studies were performed to predict the binding modes of compound **#1**, compound **#2m**, and compound **#2d** using the X-ray crystal structure of human tryptase beta-2 (hTPSB2) in its apo form. Compounds were converted to 3D structures and prepared by using Schrödinger's LigPrep tool. This process generated multiple states for stereoisomers, tautomers, ring conformations (one stable ring conformer by default), and protonation states. In particular, another Schrödinger package, Epik 6.8, was used to assign tautomers and protonation states that would be dominant at a selected pH range (pH = 7 ± 1). Ambiguous chiral centers were enumerated, allowing a maximum of 32 isomers to be produced from each input structure. Then, energy minimization was performed with the OPLS3 forcefield.

The protein was prepared using Maestro Protein Preparation Wizard. Hydrogen atoms were added, and water molecules were removed from the protein structure.

The GENEOnet 1.2 tool was employed to define protein-binding pockets and guide the docking experiments [16]. Developed by Dompé Farmaceutici SpA, this proprietary software integrates the geometric and explainability features of Group Equivariant Non-Expansive Operators (GENEOs) within a network architecture, creating a novel, knowledge-driven machine learning paradigm. By leveraging critical chemical–physical properties such as lipophilicity, hydrophilicity, and electrostatics, GENEOnet effectively identifies and prioritizes binding sites. For each parameter, a distinct GENEO [17] is employed to pinpoint regions with optimal values, enhancing the precision of pocket detection.

The docking simulations were performed by using LiGen. LiGen, a proprietary software developed by Dompé, implements a geometrical fitting procedure combined with a rigid body minimization. The Chemical Score, representing the ligand-binding interaction energy, is calculated using an in-house-developed scoring function after an initial rigid-body minimization is performed to optimize the docked ligand within the binding site. All poses that do not fulfill geometric fitting or threshold values of user-defined specific parameters are discarded.

Covalent docking was performed using the Covalent Docking 10.3 module of Schrodinger.

5. Conclusions

We have developed a novel and reliable SPR-based binding assay to evaluate potential ligands to human tryptase, measuring the underlying binding constants and stoichiometry of the ligand–protein complex. The assay proved to be a valuable tool to integrate in silico predictions and obtain important experimental insights into both the strength and the mode of interaction of ligands with the tetrameric human tryptase that was finally validated by X-ray crystal structures of one tryptase–inhibitor complex. This can provide a surrogate or additional method to investigate and study the binding properties and stoichiometry of small molecules against their target, limiting the expensive costs of structural microscopy techniques.

This integrated and multimodal approach can be used for future screening campaigns to identify new ligands of the human tryptase, levering the synergism between the in silico prediction and the experimental validation.

Author Contributions: Conceptualization, A.R.B., D.I., V.S., M.G. and M.B. (Marten Beeg); methodology, A.P., C.M., C.T., M.B. (Margherita Brindisi) and M.B. (Marten Beeg); formal analysis, A.P., C.M., C.T., D.I., M.B. (Margherita Brindisi), V.S., M.G. and M.B. (Marten Beeg); investigation, A.P., C.M., C.T., M.B. (Margherita Brindisi) and M.B. (Marten Beeg); writing—original draft preparation, A.P., C.M., D.I., M.G. and M.B. (Marten Beeg); writing—review and editing, A.P., C.M., C.T., A.R.B., D.I., M.B. (Margherita Brindisi), V.S., M.G. and M.B. (Marten Beeg). All authors have read and agreed to the published version of the manuscript.

Funding: The laboratory of Marco Gobbi has received financial support from Dompé Farmaceutici SpA for experimental analysis, including those reported in the present work.

Institutional Review Board Statement: Not applicable.

Informed Consent Statement: Not applicable.

Data Availability Statement: Data are available upon request.

Conflicts of Interest: Authors Candida Manelfi, Carmine Talarico, Andrea Rosario Beccari and Daniela Iaconis are employed by the company EXSCALATE-Dompé Farmaceutici SpA. The remaining authors declare that the research was conducted in the absence of any commercial or financial relationships that could be construed as a potential conflict of interest.

References

1. Elieh Ali Komi, D.; Wöhrl, S.; Bielory, L. Mast Cell Biology at Molecular Level: A Comprehensive Review. *Clin. Rev. Allerg. Immunol.* **2020**, *58*, 342–365. [CrossRef] [PubMed]
2. Atiakshin, D.; Buchwalow, I.; Samoilova, V.; Tiemann, M. Tryptase as a Polyfunctional Component of Mast Cells. *Histochem. Cell Biol.* **2018**, *149*, 461–477. [CrossRef] [PubMed]
3. Hallgren, J.; Pejler, G. Biology of Mast Cell Tryptase. *FEBS J.* **2006**, *273*, 1871–1895. [CrossRef] [PubMed]
4. Levi-Schaffer, F.; Piliponsky, A.M. Tryptase, a Novel Link between Allergic Inflammation and Fibrosis. *Trends Immunol.* **2003**, *24*, 158–161. [CrossRef] [PubMed]
5. Ribatti, D. Tryptase and Tumor Angiogenesis. *Front. Oncol.* **2024**, *14*, 1500482. [CrossRef] [PubMed]
6. O'Connell, M.P.; Lyons, J.J. Resolving the Genetics of Human Tryptases: Implications for Health, Disease, and Clinical Use as a Biomarker. *Curr. Opin. Allergy Clin. Immunol.* **2022**, *22*, 143. [CrossRef] [PubMed]
7. Cao, M.; Gao, Y. Mast Cell Stabilizers: From Pathogenic Roles to Targeting Therapies. *Front. Immunol.* **2024**, *15*, 1418897. [CrossRef] [PubMed]
8. Giardina, S.F.; Werner, D.S.; Pingle, M.; Bergstrom, D.E.; Arnold, L.D.; Barany, F. A Novel, Nonpeptidic, Orally Active Bivalent Inhibitor of Human β-Tryptase. *Pharmacology* **2018**, *102*, 233–243. [CrossRef] [PubMed]

9. Palmer, J.T.; Rydzewski, R.M.; Mendonca, R.V.; Sperandio, D.; Spencer, J.R.; Hirschbein, B.L.; Lohman, J.; Beltman, J.; Nguyen, M.; Liu, L. Design and Synthesis of Selective Keto-1,2,4-Oxadiazole-Based Tryptase Inhibitors. *Bioorg Med. Chem. Lett.* **2006**, *16*, 3434–3439. [CrossRef] [PubMed]
10. Schasfoort, R.B.M. *Handbook of Surface Plasmon Resonance*; The Royal Society of Chemistry: London, UK, 2017; ISBN 978-1-78262-730-2.
11. Homola, J.; Piliarik, M. *Surface Plasmon Resonance Based Sensors*; Homola, J., Ed.; Springer Series on Chemical Sensors and Biosensors; Springer: Berlin/Heidelberg, Germany, 2006; Volume 4, ISBN 978-3-540-33918-2.
12. Sofiyev, V.; Kaur, H.; Snyder, B.A.; Hogan, P.A.; Ptak, R.G.; Hwang, P.; Gochin, M. Enhanced Potency of Bivalent Small Molecule Gp41 Inhibitors. *Bioorganic Med. Chem.* **2017**, *25*, 408–420. [CrossRef] [PubMed]
13. Guo, X.-K.; Zhang, Y. CovBinderInPDB: A Structure-Based Covalent Binder Database. *J. Chem. Inf. Model.* **2022**, *62*, 6057–6068. [CrossRef] [PubMed]
14. Costanzo, M.J.; Yabut, S.C.; Zhang, H.-C.; White, K.B.; De Garavilla, L.; Wang, Y.; Minor, L.K.; Tounge, B.A.; Barnakov, A.N.; Lewandowski, F.; et al. Potent, Nonpeptide Inhibitors of Human Mast Cell Tryptase. Synthesis and Biological Evaluation of Novel Spirocyclic Piperidine Amide Derivatives. *Bioorganic Med. Chem. Lett.* **2008**, *18*, 2114–2121. [CrossRef] [PubMed]
15. Akçay, G.; Belmonte, M.A.; Aquila, B.; Chuaqui, C.; Hird, A.W.; Lamb, M.L.; Rawlins, P.B.; Su, N.; Tentarelli, S.; Grimster, N.P.; et al. Inhibition of Mcl-1 through Covalent Modification of a Noncatalytic Lysine Side Chain. *Nat. Chem. Biol.* **2016**, *12*, 931–936. [CrossRef] [PubMed]
16. Bocchi, G.; Frosini, P.; Micheletti, A.; Pedretti, A.; Gratteri, C.; Lunghini, F.; Beccari, A.R.; Talarico, C. GENEOnet: A New Machine Learning Paradigm Based on Group Equivariant Non-Expansive Operators. An Application to Protein Pocket Detection. *arXiv* **2022**, arXiv:2202.00451. [CrossRef]
17. Bergomi, M.G.; Frosini, P.; Giorgi, D.; Quercioli, N. Towards a Topological–Geometrical Theory of Group Equivariant Non-Expansive Operators for Data Analysis and Machine Learning. *Nat. Mach. Intell.* **2019**, *1*, 423–433. [CrossRef]

molecules

MDPI

Article

Identification of Novel GANT61 Analogs with Activity in Hedgehog Functional Assays and GLI1-Dependent Cancer Cells

Dina Abu Rabe [1], Lhoucine Chdid [2], David R. Lamson [2], Christopher P. Laudeman [2], Michael Tarpley [2], Naglaa Elsayed [2], Ginger R. Smith [2], Weifan Zheng [2,3], Maria S. Dixon [2,*] and Kevin P. Williams [2,3,*]

1 INBS PhD Program, North Carolina Central University, Durham, NC 27707, USA; daburabe@eagles.nccu.edu
2 Biomanufacturing Research Institute and Technology Enterprise, North Carolina Central University, Durham, NC 27707, USA; lchdid@omniab.com (L.C.); davidlamson@yahoo.com (D.R.L.); cpl@laudeman.com (C.P.L.); mtarpley@nccu.edu (M.T.); nelsayed@eagles.nccu.edu (N.E.); grsmith@nccu.edu (G.R.S.); wzheng@nccu.edu (W.Z.)
3 Department of Pharmaceutical Sciences, North Carolina Central University, Durham, NC 27707, USA
* Correspondence: maria.m.dixon37@gmail.com (M.S.D.); kpwilliams@nccu.edu (K.P.W.)

Abstract: Aberrant activation of hedgehog (Hh) signaling has been implicated in various cancers. Current FDA-approved inhibitors target the seven-transmembrane receptor Smoothened, but resistance to these drugs has been observed. It has been proposed that a more promising strategy to target this pathway is at the GLI1 transcription factor level. GANT61 was the first small molecule identified to directly suppress GLI-mediated activity; however, its development as a potential anti-cancer agent has been hindered by its modest activity and aqueous chemical instability. Our study aimed to identify novel GLI1 inhibitors. JChem searches identified fifty-two compounds similar to GANT61 and its active metabolite, GANT61-D. We combined high-throughput cell-based assays and molecular docking to evaluate these analogs. Five of the fifty-two GANT61 analogs inhibited activity in Hh-responsive C3H10T1/2 and Gli-reporter NIH3T3 cellular assays without cytotoxicity. Two of the GANT61 analogs, BAS 07019774 and Z27610715, reduced *Gli1* mRNA expression in C3H10T1/2 cells. Treatment with BAS 07019774 significantly reduced cell viability in Hh-dependent glioblastoma and lung cancer cell lines. Molecular docking indicated that BAS 07019774 is predicted to bind to the ZF4 region of GLI1, potentially interfering with its ability to bind DNA. Our findings show promise in developing more effective and potent GLI inhibitors.

Keywords: hedgehog; GLI1; GANT61; C3H10T1/2; high-throughput screening; GLI inhibitors; molecular docking

1. Introduction

GLI1 (glioma-associated oncogene homolog 1) was first identified in human glioma cancer cell lines as an oncogene [1] and subsequently shown to be a zinc finger (ZF) containing transcription factor [2,3]. GLI1 was shown to be a transcription factor that acts as the main effector of hedgehog (Hh) signaling [4,5], a major morphological pathway involved in development [6–8]. In mammals, GLI1 and its family members GLI2 and GLI3 all contribute to the transcription response to Hh activation (as recently reviewed in [6]). During canonical Hh signaling, the binding of the Hh ligand to its receptor, Patched 1 (PTCH1), relieves PTCH1 repression of Smoothened (SMO), a seven-pass transmembrane protein. This allows for the activation of SMO and the translocation of GLI proteins into the nucleus, turning on the transcription of various Hh target genes, including GLI1 itself. GLI1 activation creates a positive feedback loop, amplifying the overall Hh response [8].

Hh/GLI activation has been observed in various human tumor types [8–11]. This activation has been reported to occur through both canonical and non-canonical mechanisms, with GLI1 as the key transcription factor [12–18]). GLI1 expression has been associated with a poor prognosis and an advanced stage of various cancers [16,19]. GLI1 activation can occur through multiple pathways independent of SMO, including by PI3K/AKT/mTOR, TGFβ, the DYRK family, and oncogenic drivers such as c-Myc [15,16,20–22]. Noncanonical GLI1 activation has been reported in esophageal [23], glioblastoma [24], and lung [25] cancers. Therefore, identifying drugs that target the Hh/GLI pathway has been a major focus in cancer drug development. Small-molecule inhibitors of SMO have shown the most success in targeting the Hh/GLI pathway [26,27], resulting in the development of three FDA-approved SMO inhibitors [28–30]. Despite the initial success of SMO inhibitors, clinical studies have reported resistance to these drugs [27,31]. This has led to identifying other potential drug targets in the pathway, including downstream at the level of GLI [20,32–38].

GLI1/DNA binding has been proposed as a druggable target [39]. GANT61, discovered by small-molecule screening in a *Gli* reporter cell-based assay [40], was shown to directly block GLI-mediated transcriptional activity [40]. Subsequent molecular docking and biochemical studies showed that GANT61 binds directly to GLI1 within the zinc finger (ZF) region [41]. Additionally, GANT61 has demonstrated anti-cancer activity in various cell studies, including the suppression of cancer cell growth [42,43] and the promotion of apoptosis and G1/S phase cell cycle arrest [44]. However, GANT61 was shown to be unstable under physiological conditions and hydrolyzes into an aldehyde derivative and a diamine (GANT61-D), which is the bioactive form that inhibits GLI activity [45]. Several other GLI-directed inhibitors have been identified [20,35,37,46,47]), including GANT58 [40], HPI-1-4 [48], glabrescione B [39], arsenic trioxide [49], genistein [50], pirfenidone [51], pyrvinium [52], FN1-8 [53,54], ketoprofen [55,56], nanoquinacrine [57], zerumbone [58], 5′-O-Methyl-3-hydroxyflemingin [59], and cynanbungeigenin C and D [60]. With the exception of arsenic trioxide, which is not a GLI-specific inhibitor [49,61], these GLI antagonists have not been clinically evaluated [37,46].

In our study, we aimed to identify novel and potent GLI antagonists with potentially improved physicochemical properties. Utilizing JChem for chemical structure searches, we identified fifty-two analogs similar to both GANT61 and its active form, GANT61-D. The screening of these fifty-two analogs in high-throughput Hh functional cell-based assays identified six potential inhibitors with potencies in the low micromolar range. Our qRT-PCR data confirmed that two analogs (BAS 07019774 and Z27610715) significantly inhibited *Gli1* mRNA expression in Hh-responsive C3H10T1/2 cells. Notably, BAS 07019774 inhibited the growth of GLI1-dependent glioma and lung cancer cells. Further, a computational molecular docking simulation predicted that BAS 07019774 binds to the ZF4 domain within the GLI1-DNA binding region. The amino acid residues interacting with BAS 07019774 are conserved between GLI1 and GLI2, suggesting a broader targeting potential. The identification of BAS 07019774 as a promising GLI antagonist provides a novel starting point for future research aimed at developing improved GLI-targeted compounds.

2. Results

2.1. Generating the GANT61 Analog Compound Set

To identify compounds structurally similar to GANT61 [40], we performed the Tanimoto 3D similarity search in Instant JChem (Chemaxon Ltd., Budapest, Hungary). This first search yielded 150 potential compounds (with coefficients ranging from 0.8 down to 0.5), and after assessing their availability, 43 compounds were identified for testing. Subsequent to our initial similarity search, it was reported that the active form of GANT61 was a diamine derivative (termed GANT61-D) resulting from the hydrolysis of the parent compound [45]. Hence, a second round of similarity searching using the GANT61-D chemical scaffold in JChem was performed. This search (with a coefficient >0.5) yielded 55 potential compounds, including GANT61. Many of these compounds overlapped with the initial search results with an additional nine novel candidates identified (Figure 1A). From the combined results of both searches, 52 GANT61 analogs were purchased from vendors as listed in Supplementa Table S1.

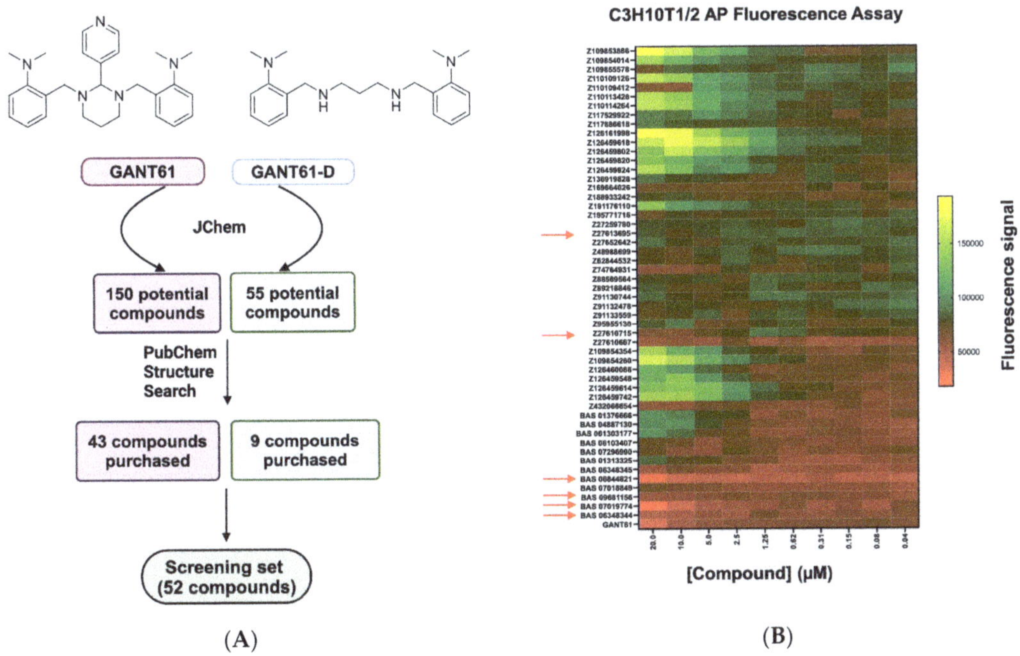

Figure 1. Screening of GANT61 analog compound set for hedgehog pathway inhibition. (**A**) Illustration of JChem and PubChem structure searches to identify compounds that share similar pharmacophores with GANT61 and the active form of GANT61 (GANT61-D). (**B**) Representative heat map profile for Hh pathway inhibitory effects of the GANT61 analog set in C3H10T1/2 cells. C3H10T1/2 cells were stimulated with SAG at its EC_{50} value (30 nM). GANT61 analogs were added in dose–response at the indicated concentrations, and after 5 days, alkaline phosphatase (AP) activity was measured using a fluorescent AP substrate. Each row represents dose–response data for a single GANT61 analog. The heat map key indicates that red stands for maximum inhibition and pale green stands for minimal inhibition. Red arrows indicate GANT61 analogs selected for further study.

2.2. Assessing GANT61 Analogs for Inhibition of Hh Activity in the C3H10T1/2 Cell-Based Assay

The GANT61 analog set was first assessed for Hh pathway inhibition using the Hh-responsive C3H10T1/2 cell line assay, a well-accepted bioassay to evaluate inhibitors for this pathway [62,63]. In these murine pluripotent cells, Hh activation leads to alkaline phosphatase (AP) induction, a marker for differentiation into osteoblast lineages [64,65]. Compounds underwent high-throughput screening using an automated and optimized 384-well format with either a fluorescence or absorbance readout for AP activity. To induce Hh pathway activity, C3H10T1/2 cells were treated with SAG at its EC_{50} concentration (30 nM) (Supplemental Figure S1A) followed by GANT61 analog compounds added in dose–response. Cells were incubated for a further 5 days, and then the AP activity was measured. This initial screening of the fifty-two GANT61 analogs identified several compounds that decreased SAG-induced AP activity in the C3H10T1/2 cells (the fluorescence (flu) readout is shown in Figure 1B and the absorbance (abs) readout in Supplemental Figure S2A). Compounds with activity in only one readout (flu or abs) were excluded to eliminate those potentially interfering with the assay [66], resulting in six compounds being selected for further analysis. These six compounds were evaluated to determine their potency in C3H10T1/2 cells with the flu AP readout. Two analogs, BAS07019774 (IC_{50} = 5.5 μM) and Z27610715 (IC_{50} = 1.1 μM), exhibited stronger inhibitory activity compared to that of GANT61 (IC_{50} = 10.5 μM) (Figure 2A, Table 1). The remaining four analogs showed incomplete inhibition and did not reach 50% inhibition at the highest concentration tested.

In the C3H10T1/2 colorimetric assay, the six GANT61 analogs showed inhibition profiles (Supplemental Figure S2B) comparable to those we observed in the fluorescence-based version of the assay (Figure 2A), with BAS07019774 and GANT61 again having full dose–response curves, yielding IC_{50} values of 3.3 μM and 3.7 μM, respectively (Table 1).

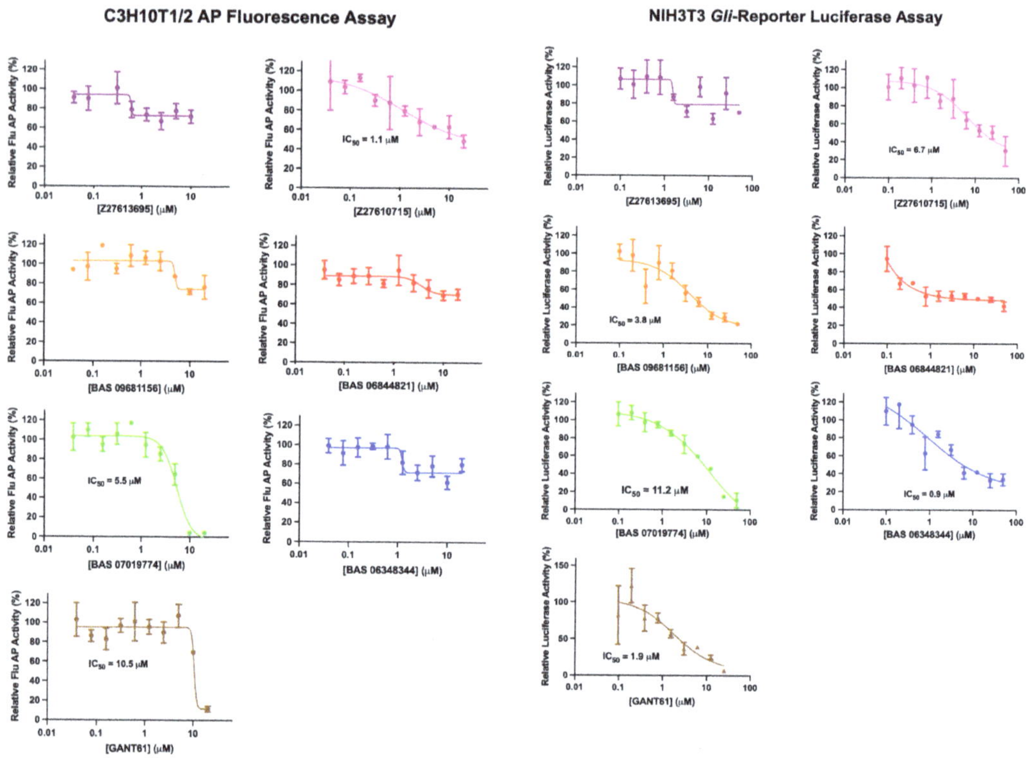

(A) (B)

Figure 2. Dose–response curves for selected GANT61 analogs in Hh-responsive cellular assays C3H10T1/2 and *Gli* reporter NIH3T3. (**A**) The C3H10T1/2 AP fluorescence assay was carried out as in Figure 1. Data were normalized to SAG-stimulated cells and are displayed as the mean of % activity (n = 3 independent experiments). (**B**) *Gli* reporter NIH3T3 cells were treated with SAG at its EC_{50} value (26 nM), GANT61 analogs added in dose–response at the indicated concentrations, and cells were incubated at 37 °C for 30 h. Firefly luciferase activity was measured as described in Methods. Data were normalized to cells treated with SAG (negative control) and are displayed as the mean of % luciferase activity ± SD (n = 3 independent experiments). Dose–response curves were generated using non-linear regression, and IC_{50} values were determined using GraphPad 9.

For further confirmation, the six GANT61 analog compounds were repurchased and retested for inhibition in the C3H10T1/2 assay using SAG and as an additional control, purmorphamine (PUR) [67,68], another Smo agonist to activate the Hh pathway (PUR, EC_{50} = 500 nM; Supplemental Figure S1A). The Hh-pathway-inhibitor KAAD-cyclopamine (KAAD-cyc) [69] was used as a positive inhibitor control (IC_{50} = 6.2 nM; Supplemental Figure S1A). The six GANT61 analogs displayed comparable inhibition profiles and IC_{50} values regardless of which Smo agonist (SAG or PUR) was used (Supplemental Figure S3). For Z27610715, BAS07019774, and GANT61, IC_{50} values of 8.1, 6.5, and 2.5 μM, respectively, were determined when using PUR as the pathway activator (Table 1).

Table 1. Activity testing in C3H10T1/2 and Gli-luciferase NIH3T3 cell-based assays for compounds identified from primary screening.

Compound	Chemical Structure	C3H10T1/2 (Flu AP, IC_{50} μM)	Gli-NIH3T3 (luciferase, IC_{50} μM)	C3H10T1/2 (Abs AP, IC_{50} μM)	C3H10T1/2 (PUR; Flu AP, IC_{50} μM)
Z27613695	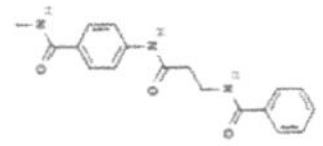	<20% [a]	<20%	<20%	<20%
Z27610715	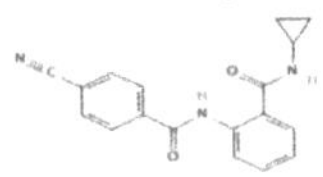	1.1	6.7	<20%	8.1
BAS 06844821	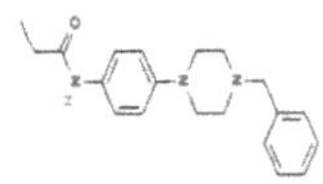	<20%	~40%	<20%	<20%
BAS 09681156	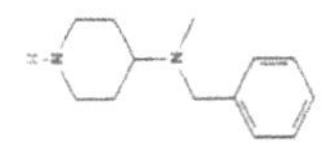	<20%	3.8	~30%	~40%
BAS 07019774	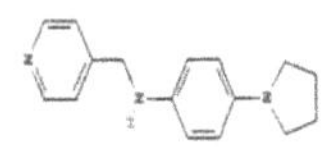	5.5	11.2	3.3	6.5
BAS 06348344	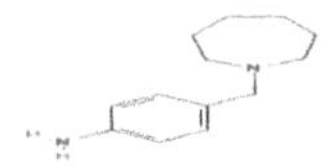	<20%	0.9	<20%	<20%
GANT61	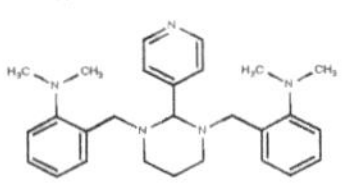	10.5	1.9	3.7	2.5

[a] For values given as %, this refers to % inhibition at highest dose tested.

2.3. Testing Selected GANT61 Analogs for Inhibition in the Gli-Luciferase Reporter NIH3T3 Cell Line

To confirm Hh pathway inhibition and to eliminate any cell-line-specific effects, we next tested the six identified GANT61 analogs in an orthogonal Hh-responsive cell line model, the Gli-luciferase reporter NIH3T3 cell line (BPS Bioscience, San Diego, CA, USA). In these engineered NIH3T3 cells, the gene for firefly luciferase was stably integrated to be under the transcriptional control of Gli-responsive elements. To assess the six GANT61 analogs, the Gli-luciferase NIH3T3 cells were stimulated with SAG at its EC_{50} concentration (26 nM, Supplemental Figure S1B), compounds added in dose–response (with KAAD-cyc as a positive control inhibitor (IC_{50} = 8.8 nM); Supplemental Figure S1B), and the luciferase activity measured. Four of the six GANT61 analogs significantly decreased the luciferase expression in a dose-dependent manner with Z27610715, BAS 09681156, BAS 07019774, and BAS 06348344 having IC_{50} values of 6.7, 3.8, 11.2, and 0.9 μM, respectively, and GANT61 having an IC_{50} value of 1.9 μM (Figure 2B, Table 1). BAS 06844821 had more modest effects on the luciferase activity, decreasing the activity by ~40% at the highest dose tested. As Z27613695 showed no significant inhibition of activity in any of the Hh assays (Table 1), this compound was no longer included for subsequent testing. The results from both of the Hh-responsive assays showed BAS 07019774 and Z27610715 were the most active (Table 1).

2.4. Assessing Selected GANT61 Analogs for Cell Cytotoxicity in the Hh-Cell-Based Assays

Next, we assessed whether the GANT61 analogs specifically inhibited Hh signaling or caused general cytotoxicity in the C3H10T1/2 and Gli reporter NIH3T3 cells. These cells were stimulated with SAG, selected GANT61 analogs added in dose–response, and the effects on cell number determined using the nuclear stain Hoechst 33342 [70]. The GANT61 analogs did not significantly affect the C3H10T1/2 cell number, while GANT61 significantly decreased the cell number at the highest concentration of 20 μM (a 93.5% reduction) (Figure 3A). Likewise, no significant reduction in Gli reporter NIH3T3 cell numbers was observed with the GANT61 analogs tested up to a concentration of 50 μM (Figure 3B). However, GANT61 showed significant cytotoxicity in these cells with cell count reductions of 49.4% and 86.5% at 25 μM and 50 μM, respectively (Figure 3B). These findings suggest that the inhibitory effects of the GANT61 analogs in both Hh-responsive cell lines are mediated through the Hh pathway rather than by a general cytotoxic effect.

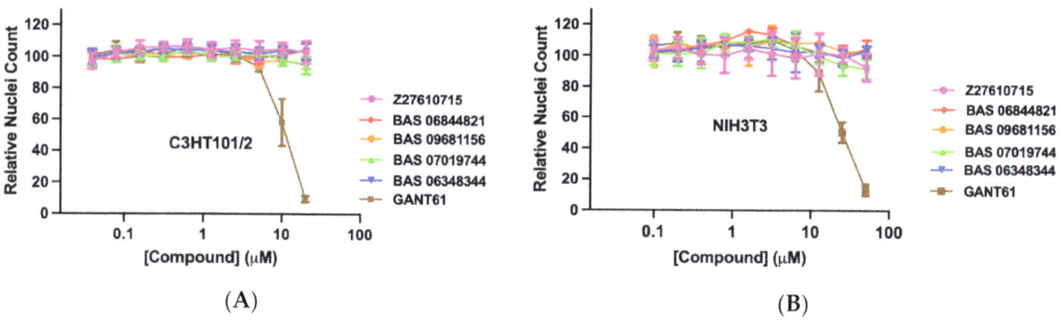

Figure 3. Effect of the selected GANT61 analogs on cell cytotoxicity in C3H10T1/2 and *Gli* reporter NIH3T3 cells. GANT61 analogs were added in dose–response at the indicated concentrations to C3H10T1/2 cells for 5 days (**A**) or to *Gli* reporter NIH3T3 cells for 30 h (**B**). Cells were then incubated with Hoechest-33342, and cell numbers were determined using high-content imaging as described in Methods. Data were normalized to SAG-treated cells. Data are displayed as the mean of the relative nuclei count ± SD (n = 3 independent experiments).

2.5. Testing GANT61 Analogs for Inhibition of SAG-Induced Gli1 mRNA Expression in C3H10T1/2 Cells

To determine the direct inhibition of Hh pathway transcriptional activity by the GANT61 analogs, Gli1 mRNA expression was measured in SAG-induced C3H10T1/2 cells. A preliminary time course assessed Gli1 mRNA expression in C3H10T1/2 cells induced by SAG, with two time points selected for further studies (48 and 96 h, Supplemental Figure S4A). Based on their consistent inhibitory activity in the previous Hh-responsive assays (Table 1), we selected the GANT61 analogs Z27610715 and BAS 0701977 for continued testing. C3H10T1/2 cells were stimulated with SAG at its EC_{50} concentration (30 nM) and treated with the two GANT61 analogs, and the Gli1 mRNA levels were measured using TaqMan qRT-PCR. As expected, SAG significantly increased Gli1 mRNA expression at 48 h (27-fold) and 96 h (278-fold) in comparison to that of untreated cells (Figure 4A,B). The positive control inhibitor, KAAD-cyc, effectively inhibited SAG-induced Gli1 mRNA expression (with a six-fold decrease at 48 h; Supplemental Figure S4B). BAS 07019774 significantly reduced Gli1 mRNA expression at both time points (4.9-fold at 48 h and 2.2-fold at 96 h) compared to the SAG-treated cells (Figure 4A,B), while Z27610715 showed a significant reduction only at the later time point (4.5-fold at 96 h) (Figure 4B). Interestingly, GANT61 reduced Gli1 mRNA expression at 48 h (20.4-fold) but did not significantly reduce Gli1 mRNA expression at 96 h (Figure 4B). These findings further support the inhibitory activity of the GANT61 analogs on the Hh pathway.

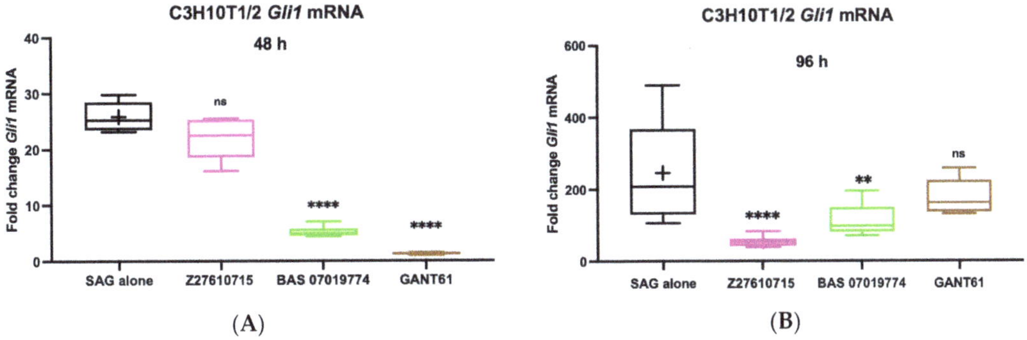

Figure 4. Effect of GANT61 analogs on *Gli1* mRNA expression in C3H10T1/2 cells. C3H10T1/2 cells were seeded in 12-well plates at a cell density of 60,000 cells/well, treated with SAG (30 nM) alone (control) or with SAG in the presence of 10 μM of BAS07019774, Z27610715, or GANT61. RNA was collected at 48 (**A**) and 96 (**B**) h, and *Gli1* mRNA expression was determined by Taqman qRT-PCR. Data were normalized to the housekeeping gene β-actin. The mean of three independent experiments ± SD is shown. Data were evaluated by one-way ANOVA followed by Tukey's multiple comparisons test using GraphPad Prism 9. Significant differences relative to treatment of SAG alone (**** $p < 0.0001$, ** p-value = 0.0022, ns = not significant).

2.6. Effects of GANT61 Analogs on the Viability of Glioblastoma and Lung Cancer Cell Lines

To evaluate the effect of the identified GANT61 analogs on cancer cell viability, we first assessed the expression of GLI1 and GLI2 in two human glioblastoma cell lines, U87MG and T98G, previously shown to be GANT61-sensitive and GLI1-dependent [71–73]. Consistent with previous reports [71,73], we observed that U87MG cells express both GLI1 and GLI2 and that T98G expresses predominantly GLI1 (Figure 5A and Supplemental Figure S5). The GLI1 mRNA expression was comparable in both U87MG and T98G (Figure 5A, Supplemental Figure S5). U87MG and T98G cells were incubated with GANT61 analogs for 72 h, and the cell viability was measured. Notably, BAS 07019774 showed significant effects on both U87MG and T98G cell viability (with IC_{50} values of 9.5 and 29.5 μM, respectively) (Figure 5B). Z27610715 exhibited a moderate reduction in T98G cell viability (a 19.2% reduction at 50 μM). Further, BAS 07019774 was more effective in reducing U87MG cell viability in comparison to GANT61, while their effects on TG98 cells were comparable (Figure 5B).

We further explored the effects of BAS 07019774 on two lung cancer cell lines, SK-MES-1 and H1437. SK-MES-1 has been shown to be GLI-dependent and GANT61-sensitive [74]. We found that the SK-MES-1 cells expressed both GLI1 and GLI2, which is consistent with a previous report [74], while the H1437 cells expressed minimal levels of GLI1 and GLI2 (Figure 5A and Supplemental Figure S5). Our data showed that BAS 07019774 was able to reduce the SK-MES-1 cell viability (with an IC_{50} value of 9.3 μM), whereas GANT61 had a modest effect (Figure 5C). Significantly, BAS 07019774 did not affect the viability of H1437 cells that have minimal GLI expression, while GANT61 had some effect on the viability of these cells at high concentrations (Figure 5C).

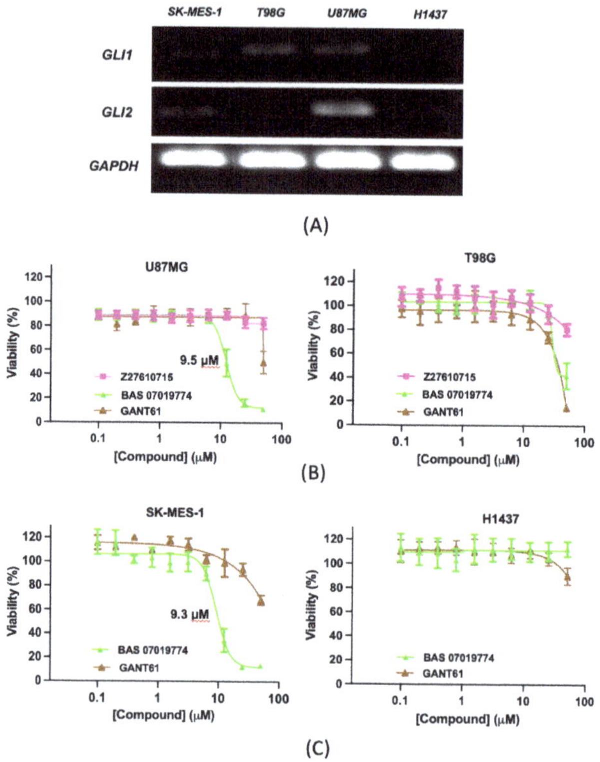

Figure 5. Effect of the selected GANT61 analogs on glioblastoma and lung cancer cell viability. (**A**) *GLI1* and *GLI2* mRNA expression levels were quantified in human glioblastoma and lung cancer cell lines (U87MG, T98G, SK-MES-1, and H1437) using TagMan qRT-PCR as described in Methods (see data in Supplemental Figure S5). PCR products were electrophoresed on 2% agarose gels and visualized by iBright 1500 Imaging System (Invitrogen, Carlsbad, CA, USA). The bands obtained for *GLI1*, *GLI2*, and *GAPDH* are at the predicted sizes of 80 bp, 88 bp, and 157 bp, respectively. Glioblastoma (**B**) and lung cancer (**C**) cell viability using MTT assay. Glioblastoma U87MG and T98G cells were seeded in 384-well plates at densities of 2500 and 1500 cells/well, respectively. SK-MES-1 and H1437 lung cancer cells were seeded in 384-well plates at densities of 1000 cells/well and 2000 cells/well, respectively. Cells were treated with the GANT61 analogs in dose–response at the indicated concentrations for 72 h, and their viability was assessed by MTT assay as described in Methods. Data are presented as the mean of % viability relative to that of cells treated with vehicle (0.1% DMSO). Dose–response curves were generated using non-linear regression, and IC_{50} values were determined using GraphPad Prism 9.

2.7. Predicted Binding Mode of BAS 07019774

GANT61 was previously predicted to bind at the ZF region of GLI1 [41,45]. To understand how BAS 07019774 might interact with GLI1, we performed molecular docking simulations. Molecular docking was carried out using Molecular Operating Environment (MOE) and the human GLI1-ZF/DNA crystal structure (PDB ID: 2GLI) [75]. The 3D structures of BAS 07019774 and GANT61-D were generated using energy minimization in MOE. An evaluation of the 12 lowest-energy poses (visually, the number of key interactions; Supplementa Table S2) suggested pose 7 as the most favorable binding mode for BAS 07019774. In this predicted pose, BAS 07019774 interacts with ZF4, forming three interactions, one hydrogen (H)-bond with Arg348 and two π-acceptor H-bonds [76] with His335 and Glu334 (Figure 6A–C). These interacting residues are positioned near the GLI1-ZF DNA binding site (Figure 6D) and are also conserved between GLI1 and GLI2 proteins (Figure 6D). MOE predicted that GANT61-D also interacted in the same region of GLI1-ZF, forming two H-bonds, one with Arg348 and one with Glu334 (Supplemental Figure S6).

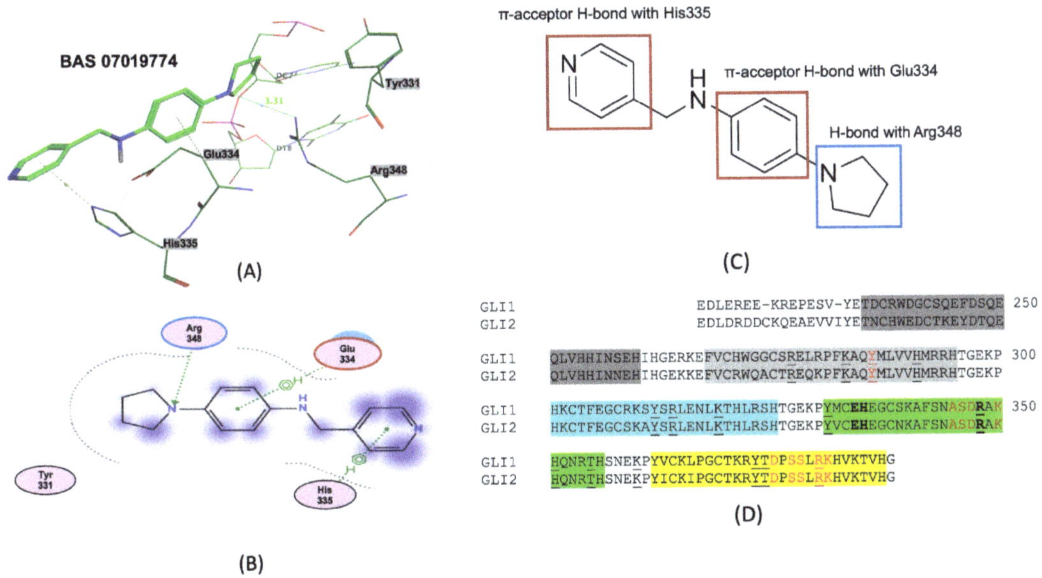

Figure 6. Predicted docking pose of BAS 07019774 to GLI1 zinc finger structure. BAS 07019774 was docked to the crystal structure of GLI1 five-zinc finger domain (PDB ID: 2GLI) using MOE. For clarity, residue numbering used in this figure corresponds to full-length GLI1 numbering (UNIPROT P08151). (**A**) The best docking pose predicted binding of BAS 07019774 to Arg348 (one hydrogen bond), Glu334 (π-acceptor H-bond), and His335(π-acceptor H-bond) residues within GLI1 ZF4. (**B**) Compounds and residues are shown in stick representation. Two-dimensional ligand interaction diagram of GLI1-ZF interactions with BAS 07019774. Ligand interactions were generated using MOE software (version MOE2022.02). (**C**) Chemical structure of BAS 07019774 annotated with GLI1-ZF interactions. (**D**) Amino acid sequence alignment of GLI1-ZF and GLI2-ZF domains is highlighted to show the following individual ZF sections: ZF1 (gray), ZF2 (light gray), ZF3 (blue), ZF4 (green), and ZF5 (yellow). Residues in bold (E, H, and R) correspond to Glu334, His335, and Arg348 and show where BAS 07019774 is predicted to bind. Residues in red are those that interact with DNA, and underlined residues are DNA phosphate contacts [75].

3. Discussion

As the hedgehog pathway has a major role in tumorigenesis, significant efforts have been made to develop drugs targeting this pathway [26,27], with several SMO inhibitors having been approved by the FDA [28–30]. Due to drug resistance to these SMO inhibitors [31], targeting elsewhere in the pathway, including downstream at the level of GLI, has been proposed [20,32–38]. Studies indicate that the inhibition of GLI may be more effective than SMO in blocking tumor growth in several cancer models [77–79]. GANT61 was one of first compounds shown to directly inhibit GLI1/2-mediated transcription [40] and was subsequently shown to bind directly to GLI1 [41]. However, the chemical instability and poor pharmacokinetic properties of GANT61 have prevented its development as an anti-cancer agent [20,45].

In this study, GANT61 and its active form GANT61-D were used as scaffolds to identify closely related compounds that may have improved potency and drug-like properties. From a set of fifty-two GANT61 analogs, we initially identified six with inhibitory effects (a low micromolar range) on the Hh pathway in cell-based assays. Five of these six GANT61 analogs were confirmed to be active in an orthogonal Hh-responsive *Gli* reporter cell-based assay. In C3HT101/2 cells, one of the compounds, BAS 07019774 (a *p*-phenylenediamine), significantly reduced *Gli1* mRNA at both 48 and 96 h. In contrast, Z27610715 was ineffective at 48 h but potent at 96 h, while GANT61 appeared to be more effective in inhibiting *Gli1* mRNA expression at 48 h rather than at 96 h (Figure 4), suggesting differing stabilities or solubilities amongst these compounds. The higher efficacy of GANT61 analogs in reducing *Gli1* expression compared to GANT61 at 96 h may be attributed, at least in part, to the reported low level of stability of GANT61 [45]. Further studies are required to assess the stability and the solubility of the selected GANT61 analogs (BAS 07019774 and Z27610715) compared to GANT61. Further, we endeavored to address some of the challenges outlined by Curran [80] in developing Hh pathway inhibitors, in particular, their potential off-target/cytotoxic effects at higher doses. We demonstrated that BAS 07019774 was effective at low micromolar concentrations in the Hh functional cellular assays without any significant cell cytotoxicity, indicating that the inhibition by BAS 07019774 was not simply due to cell killing independent of the Hh pathway. In contrast, high concentrations of GANT61 did appear to be cytotoxic for these cells. GANT61 has been reported to have differing cytotoxic effects on "normal" cell lines (see, for example, [81]).

In GLI-dependent cancer cell models, treatment with BAS 07019774 significantly reduced cell viability in both U87MG and T98G glioblastoma cell lines. Z27610715 showed a minimal effect on U87MG cells and a modest reduction in the viability of T98G cells. While GANT61 was less effective than BAS 07019774 in U87MG, they had comparable effects in T98G cells. Notably, BAS 07019774 was more effective in reducing the cell growth of U87MG, which expresses both GLI1 and GLI2, suggesting that BAS 07019774 potentially targets both GLI1 and GLI2. To provide more evidence for GLI targeting by this compound, BAS 07019774 was also found to be effective in reducing the viability of a GLI-dependent lung cancer cell line SK-MES-1 but was ineffective in a lung cancer line with minimal GLI expression (H1437). These findings suggest that the anti-cancer activity of BAS 07019774 could be linked to an ability to target GLI proteins.

Our evaluation of the molecular docking prediction of BAS 07019774 binding to the human GLI1-ZF domain [75] showed the best pose consisted of three interactions: one hydrogen (H)-bond with Arg348 and two π-acceptor H-bond interactions with the hydrogens of His335 and Glu334. These three residues are found within ZF4 of GLI1 (Figure 6). Notably, Arg348 is located between key residues in ZF4 (Figure 6D; Ala345, Ser346, Asp347, and Lys350), which are in contact with the conserved nine-base pair DNA binding site [3,75]. Furthermore, the crystal structure of GLI1-ZF showed that the Arg348 side chain forms an H-bond with the phosphodiester oxygen of the DNA backbone [75], suggesting that BAS 07019774 may disrupt GLI1-DNA binding. Interestingly, the residues interacting with BAS 07019774 (Arg348, His335, and Glu334 in GLI1) are conserved between GLI1 and GLI2 [41,75] (Figure 6D). This conservation suggests that BAS 07019774 might interact similarly with both GLI1 and GLI2. This could explain the observed higher effectiveness of BAS 07019774 in reducing the viability of cancer cell lines expressing both GLI1 and GLI2 (U87MG and SK-MES-1) compared to T98G cells, which primarily express GLI1. Our molecular docking predicts GANT61-D docking at a site on ZF4 close to or overlapping with that of BAS 07019774, with GANT61-D forming two H-bonds: one with Arg348 and one with Glu334. However, previous docking studies using AutoDock have the neutral form of GANT61-D docking between ZF2 and ZF3 with H-bond contacts with Glu250 and Glu298 [41,45,82,83]. The mutation of these residues reduced GANT61-GLI1 binding as assessed by surface plasmon resonance and reduced GANT61-mediated Hh-pathway inhibition [41]. In the study by Calcaterra et al. [45], it was proposed that the di-protonated form of GANT61-D would be the prevalent form under physiological conditions. Their docking studies had this di-protonated form of GANT61-D docking within ZF1 and ZF2 at a negatively charged surface with its interaction driven predominantly by electrostatic forces [45]. While the discrepancy between our data and those of the previous reports may be due to using different docking approaches, including, for example, the program utilized, initial pocket identification, grid density, and potential biases in scoring functions, there is precedent for some of the other identified GLI binders to potentially interact at different sites on GLI-ZF (see the discussion below on other identified GLI binders). Further, docking with transcription factors such as GLI1 can be challenging due to their flexibility, the absence of well-defined binding pockets, and their generally flat surfaces [39]. Further studies will be required to determine the functional significance of the residues on GLI1 identified by our study and predicted to be involved in BAS 07019774 binding.

Based on the calculated chemical properties (Table 2), BAS 07019774 (pyridin-4-ylmethyl-(4-pyrrolidin-1-yl-phenyl)-amine) meets the criteria of Lipinski's rule of five (mol. wt. < 500, LogP < 5, HBD < 5, HBA < 10) [84,85]. Further, its molecular weight is somewhat less than that for both GANT61 and GANT61-D. Due to the relatively low number of analogs we screened, exploring a structure–activity relationship (SAR) for BAS 07019774 is challenging. However, we did identify two closely related but inactive analogs (BAS 06103407 and BAS 07018849) that maintain the *p*-diaminobenzene core but are missing the pyrrolidine ring (Figure 7). In our docking study, this pyrrolidine ring makes an H-bond with Arg348. Further, the basicity of the pyrrolidine nitrogen at physiological pH is not very high, and it is likely to not be fully protonated. The low molecular weight for BAS 07019774 provides opportunities to add substituents to increase potency and improve physicochemical properties without exceeding 500 Da. BAS 07019774 has a "linear" arrangement of amine centers that nonetheless allows for medicinal chemistry exploration to optimize its properties. The *p*-diaminobenzene core could be maintained with substitutions at any position or the addition of other heteroatoms to the rings in order to find analogs. An alternative would be to preserve the pyrrolidino-phenylenediemaine structure and interrogate the SAR around that moiety. The pyrrolidine ring of BAS 07019774 could also be opened or modified without negatively impacting any of the interactions suggested in this work.

Table 2. Calculated properties for BAS 07019774 and GANT61.

Compound	Mol. wt. (g/mol)	cLogP	LogS	PSA	RB	HBD	HBA
GANT61	429.6	4.9	−5.9	25	7	0	5
GANT61-D	340.5	3.9	−4.3	31	10	2	4
BAS 07019774	253.3	1.9	−3.7	28	4	1	3

Abbreviations: cLogP, calculated lipophilicity of partition coefficient; LogS, calculated aqueous solubility; PSA, polar surface area; RB, number of rotational bonds; HBD, number of hydrogen bond donors; HBA, number of hydrogen bond acceptors. Properties calculated in ChemDraw 23 (Revvity).

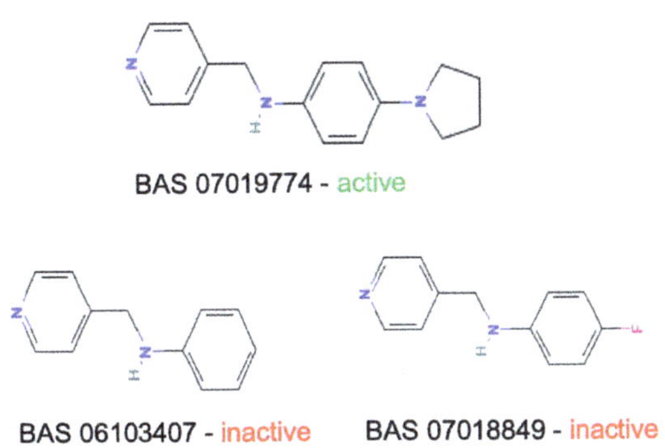

Figure 7. Preliminary SAR of BAS 07019774. Two compounds (BAS 06103407 and BAS 07018849) that have similar structures to BAS 07019774 were identified among the fifty-two analogs screened and found to be inactive. Both compounds maintain the *p*-diaminobenzene core but are missing the pyrrolidine ring, which makes an H-Bond with Arg348.

Only GANT61 [41,45], the natural isoflavone Glabrescione B (GlaB) (Figure 8) [39], and arsenic trioxide [49] have been shown to directly bind GLI1. GlaB was identified from a virtual screening of >800 compounds based on mutagenesis studies identifying residues involved at the GLI1-DNA binding site [39]. The docking studies predicted GlaB binding to a groove between ZF4 and ZF5 with interactions to Lys340 and Lys350, interfering with the interaction of GLI1 with its target DNA [39]. GlaB derivatives have been designed that can target both GLI1 and SMO [86–88]. The chemical structures of GANT61 [82,89] and GlaB [90] have been used as scaffolds for virtual screening to identity other compounds that act as GLI1 inhibitors. In particular, GlaB and vismione E (Figure 8), which were previously discovered by Infante et al. [39], were used to generate a multi-feature pharmacophore that identified thiophene and pyrazolo-pyrimidine compounds that were predicted to dock in the same binding pocket as GlaB [90]. One of the pyrazolo-pyrimidine compounds (SST0704, Figure 8) had predicted docking interactions with ZF4 [90]. This same five-feature pharmacophore was subsequently used to identify several 8-hydoxyquinoline derivatives [91] that had a similar scaffold to another 8-hydoxyquinoline analog 74 (pyridin-4-ylmethyl-(4-pyrrolidin-1-yl-phenyl)-amine) meets the criteria of L (Figure 8) [82], which was identified by virtual screening based on biased docking focused on the GANT61 binding site within ZF2 and ZF3. Interestingly, in the same study [91], a compound that matched the five-feature model (compound **1**, Figure 8) could be docked in two alternative best-docking poses: one at ZF4, as predicted for GlaB, and the other between ZF1 and ZF3, as predicted for GANT61. Further, they showed that some of the derivatives they identified could be docked at ZF4, while others also had an alternative best-scoring docking pose with the putative GANT61 binding site. Docking studies of a new 8-hydoxyquinoline derivative JC19 (Figure 8) [92] using putative binding sites at ZFs 1, 3, and 4 showed interactions at ZF4/5 similar to those hypothesized previously for other derivatives [90], as well as GlaB [39]. JC19 inhibits the formation of the GLI1-DNA complex by making predicted interactions with residues in ZF4 and ZF5, in particular, His351 and His356 [92]. These studies highlight some of the challenges with docking simulations for GLI1 and suggest that there are some pharmacophores that can be docked at both sites ("GlaB" and "GANT61"). Indeed, a recent study shows GANT61 docking with other Hh pathway components' SMO and SUFU [93].

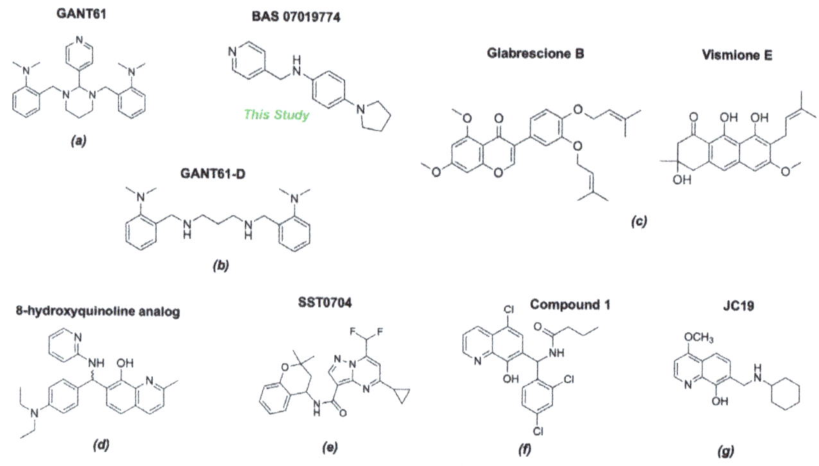

Figure 8. Chemical structures of reported direct GLI1 binding inhibitors. Structures were drawn in "ACS 1996" format with ChemDraw exchange file type (.cdx) or generated from SMILES files downloaded from PubChem. Compounds from the following references: (**a**) [40], (**b**) [45], (**c**) [39], (**d**) [82], (**e**) [90], (**f**) [91] and (**g**) [92].

While our unbiased docking for BAS 07019774 on GLI1 ZF predicted a significant cluster of the top-ranked poses at ZF4, with a predicted best pose fitting with our preliminary SAR data, we have also found that using biased docking at the GANT61 site (with a grid at Glu250/Glu298 (full-length GLI1 numbering)) can generate docking poses that have some reasonable interactions with BAS 07019774 (see an example in Supplementa Table S3). However, none of the top 15 predicted poses from this biased docking study have interactions with the pyrrolidine ring of BAS 07019774 and thus do not fit with our SAR. While some of these other GLI-ZF binding compounds (Figure 8) could bind in a similar fashion to BAS 07019774, there is not much chemical structural overlap. However, the 8-hydoxyquinoline analog ((*d*) in Figure 8) [82] is an exception that can be mapped almost directly to BAS 07019774. Further exploration around both scaffolds is necessary to establish any correlations and provide valuable insights.

4. Materials and Methods

4.1. Cell Lines, Reagents, and Compounds

Cell lines C3H10T1/2 (Clone 8), U87MG, T98G, SK-MES-1, and H1437 were obtained from American Type Culture Collection (ATCC, Manassas, VA, USA). The *Gli*-luciferase reporter NIH3T3 cell line was obtained from BPS Bioscience (San Diego, CA, USA). Cells were cultured following the manufacturer's instructions. In the case of C3H10T1/2, they were optimized for high-throughput screening (HTS) using DMEM. All experiments were conducted using cells between passage numbers 3 and 8. GANT61 was purchased from TOCRIS (Minneapolis, MN, USA). KAAD-cyclopamine (KAAD-cyc) and SAG were obtained from MilliporeSigma (Burlington, MA, USA). Purmorphamine was purchased from Cayman Chemical (Ann Arbor, MI, USA). Hoechst-33342 dye, SYPR Safe DNA gel stain, and agarose (genetic technology grade) were purchased from Themo Fisher Scientific (Waltham, MA, USA). GANT61 analogs were purchased from Enamine (Kyiv, Ukraine) or Asinex Corp (Winston-Salem, NC, USA). All compounds were purchased as powders, dissolved in 100% DMSO to obtain 10 mM and 50 mM stock solutions, and stored at −20 °C.

4.2. C3H10T1/2 Hedgehog-Responsive High-Throughput Cell-Based Assay

To assess effects on Hh activity, we used the murine embryonic fibroblast cell line C3H10T1/2 [64]. These cells have been previously demonstrated by our group and others to be responsive to Hh signaling [63,65]. C3H10T1/2 cells were cultured in Dulbecco's Modified Eagle's Medium (DMEM) supplemented with 10% FBS and plated in black, clear-bottomed 384-well plates (Thermo Fisher Scientific, Waltham, MA, USA) with 1500 cells/well using a MultiFlo cell dispenser (Agilent BioTek, Winooski, VT, USA). After 24 h, cells were treatedw with Smoothened agonist (SAG) [94,95] at its half-maximal effective concentration (EC_{50} = 30 nM) to induce Hh activity. GANT61 analogs in 100% DMSO were then added to the cell plates in a 10-point 2-fold dose–response format, and normalization performed using a D300 digital dispenser (HP). Cells were incubated for 5 days and lysed, and alkaline phosphatase (AP) activity was measured to assess Hh activity. The AP fluorescent substrate [96] (AttoPhos®, Promega, Madison, WI, USA) was added to each well, plates were incubated in the dark for 30 min, and fluorescence measured at Ex/Em 435/555 nM using a CLARIOstar plate reader (BMG Labtech, Cary, NC, USA). This automated 384-well C3H10T1/2 assay with fluorescence readout was adapted from our previous 96-well colorimetric version [97,98]. For some C3H10T1/2 cell experiments, a 384-well colorimetric readout assay (pNPP substrate, 405 nM absorbance, clear plates) was used. Purmorphamine (PUR) [67,68], an alternative SMO agonist, was used in some experiments to activate Hh activity in C3H10T1/2 cells (PUR used at its EC_{50} value of 500 nM, determined by dose–response). AP activity was normalized to SAG-induced or PUR-induced cells (maximum signal control, columns 1 and 2 in each 384-well assay plate). As a minimum signal control, cells were treated with 30 nM of SAG plus 40 nM (IC_{90}) of known Hh inhibitor KAAD-cyclopamine (KAAD-cyc) [69] (columns 23 and 24 in each

384-well assay plate). IC_{50} values were calculated using variable slope (four-parameter dose response) in Prism 9 GraphPad software.

4.3. Gli-Luciferase NIH3T3 Reporter Assay

Gli reporter NIH3T3 cells (BPS Bioscience, catalog# 60409), a stable cell line that contains firefly luciferase gene under the control of Gli-responsive elements, were used to measure Hh pathway activity. Cells were grown in 96-well plates at a density of 25,000 cells/well for 24 h in Thaw Medium 5 (BPS Bioscience). After 24 h, the medium was replaced with fresh assay media (100 μL) using a Biomek®NX (Brea, CA, USA) automated liquid handler. To activate the Hh pathway, cells were treated with SAG at its half-maximal effective concentration (EC_{50} = 26 nM). To assess inhibition, the GANT61 analogs were added (serial dilutions of 0.098 to 50 μM) using a D300 system, and cells were then incubated at 37 °C for 30 h. Firefly luciferase activity was measured using the One-Step Luciferase Assay, according to vendor's protocol (BPS Bioscience, catalog # 60690-1). Briefly, One-Step luciferase reagent (100 μL) was added to each well and incubated by shaking for 20 min at room temperature, followed by luminescence detection using a GloMax plate reader (Promega). To account for background luminescence, cell-free control wells were included. The background-subtracted luminescence of cells stimulated with SAG alone was set as 100% luminescence for calculating percent inhibition of the test compounds. The inhibition curves were plotted using GraphPad Prism 9 to determine IC_{50} values. Each assay was repeated at least three times.

4.4. Cytotoxicity Assay

Cytotoxicity of compounds was evaluated in C3H10T1/2 and *Gli* reporter NIH3T3 cell lines. Cells were plated and treated with SAG (at its EC_{50} concentration), and compounds were added in dose–response as above. For C3H10T1/2 cells, Hoechst dye 33342 (10 μg/mL) was added to each well 5 d after treatment. For *Gli* reporter NIH3T3 cells, Hoechst dye was added 24 h after treatment. Cells were incubated with the Hoechst dye for 45 min at 37 °C. Following incubation, cells were washed with (50 μL) PBS (phosphate-buffered saline), fixed with formalin (20 μL), and incubated in the dark at room temperature for 15 min. After fixing the cells, formalin was removed, and 50 μL of PBS added. High-content imaging was performed using a Cellinsight NXT system (Thermo Fisher) at an excitation wavelength of 386 nM. Data were analyzed using GraphPad Prism 9.

4.5. qRT-PCR Assay to Measure GLI mRNA Expression

Gli1 mRNA expression in murine C3H10T1/2 was quantitatively assessed by qRT-PCR, as previously described [96]. Briefly, C3H10T1/2 cells were seeded at 60,000 cells/well in 12-well plates in DMEM/10% FBS. After 24 h, cells were treated with SAG (EC_{50} = 30 nM) in the presence of GANT61 or GANT61 analogs (10 μM). Cells were harvested 48- and 96-h post-treatment (with three replicate treatments per group). *Gli1* mRNA induction by SAG over a time course (0, 12, 24, 48, 72, 96, and 120 h) was assessed in a separate experiment. Untreated cells were used as a control. Total RNA was isolated using RNeasy Mini Plus Kit (Qiagen, Germantown, MD, USA). RNA (1 μg) was reverse-transcribed into cDNA using an iScript cDNA Synthesis Kit (Bio-RAD, Hercules, CA, USA). qRT-PCR was performed to quantify mRNA using TaqMan assay. Probes specific for murine *Gli1* (Mm00494654_m1) and murine *β-actin* (Mmo2619580_g1) were obtained from Applied Biosystems (Thermo Fisher Scientific). Target sequences were amplified at 95 °C for 10 min followed by 40 cycles at 95 °C for 15 s and 60 °C for 1 min. Each experiment was performed in triplicate on a QuantStudio 6 flex (Thermo Fisher Scientific). For normalization, *β-actin* was utilized as an endogenous control, and the fold change was determined using the $2^{-\triangle\triangle Ct}$ method.

GLI1 and *GLI2* mRNA expression were quantified in human cancer cell lines (U87MG, T98G, SK-MES-1 and H1437) using TaqMan qRT-PCR as described above. Probes specific for human *GLI1* (Hs00171790_m1), *GLI2* (Hs01122187_m1), and *GAPDH* (Hs02786624_g1) were used to amplify the target sequences. The housekeeping gene GAPDH was used for normalization and to determine fold change using the $2^{-\triangle\triangle Ct}$ method. For gel electrophoresis, PCR products were run on agarose gels (2%) for 90 min at a constant 150 V. Gels were visualized with Syber Safe DNA stain and imaged using an iBright 1500 Imaging System (Invitrogen).

4.6. Cancer Cell Viability Assay

Glioma cell lines T98G and U87MG were plated in 384-well (clear, flat-bottomed) plates in EMEM/10% FBS (50 µL) at densities of 1500 and 2500 cells/well, respectively. Lung cancer cell lines SK-MES-1 and H1437 were plated at cell densities of 1000 cells/well in 50 µL EMEM and 2000 cells/well in 50 µL RPMI-1640, respectively. After 24 h, cells were treated with selected GANT61 analog compounds using a 10-point, 2-fold dose–response curve (0.098–50 µM) and incubated at 37 °C for 72 h. To assess cell viability, MTT was added to each well at a final concentration of 0.5 mg/mL using a Biomek®NX (Brea, CA, USA) automated liquid handler followed by incubation for 4 h at 37 °C. Media and MTT were then removed, DMSO (40 µL) was added to each well to dissolve the formed formazan crystals, and they were incubated for 1 h. Absorbance was then measured at 550 nM using a SpectraMax plate reader (Molecular Devices, San Jose, CA, USA). Cells treated with 0.1% DMSO were used as a control. The absorbance values from compound-treated wells were normalized to the control. The normalized absorbance values were used to calculate the percentage of viable cells remaining after treatment with the compounds. Data were analyzed using Prism 9 (GraphPad, Boston, MA, USA).

4.7. Molecular Docking

The crystallographic structure of human GLI1-ZF/DNA complex [75] (PDB ID: 2GLI) was utilized as a rigid structure to perform molecular docking using the Molecular Operating Environment (Molecular Operating Environment (MOE, version MOE2022.02), Chemical Computing Group ULC, Montreal, QC, Canada) to determine the binding mode of the identified GANT61 analogs and GANT61-D. All solvent molecules were removed from the structure prior to docking. Unbiased molecular docking simulations were carried out using MOE's default parameters. The chemical structure of BAS 07019774 was obtained from eMolecules (emolecules.com, accessed on 11 August 2023), and GANT61-D structure was as published previously [45]. Three-dimensional structure poses were generated in MOE software by energy minimization. The top-ranking poses were evaluated by comparing the docking scores (predicted binding affinity) and the number of key interactions formed.

4.8. Statistical Analysis

All cell culture and in vitro experiments were independently repeated at least three times under the same conditions. All assays were performed in triplicate. All data are presented as the mean ± SD. One-way analysis of variance (ANOVA) followed by Tukey's HSD multiple comparisons, *t*-test, and Mann–Whitney test were used to evaluate the significant differences between treatments. GraphPad Prism 9 software (San Diego, CA, USA) was used for statistical analysis. A value of $p < 0.05$ was considered statistically significant.

5. Conclusions

In conclusion, this study aimed to identify more effective Hh pathway inhibitors that target GLI. In contrast to virtual screening approaches, we utilized lab-based high-throughput technologies to successfully screen a compound library of fifty-two analogs that we identified based on the chemical structures of GANT61 and GANT61-D. Five promising analogs inhibited Hh activity in two independent cell-based assays without significant cytotoxicity. Notably, one analog, BAS 07019774, effectively decreased *Gli1* mRNA expression in an Hh-responsive cell line model and reduced cell viability in GLI-dependent cancer cell lines. Future studies will be required to evaluate the in vivo efficacy and pharmacokinetic properties of BAS 07019774. The data presented in this study provide a valuable foundation to facilitate medicinal chemistry studies to elucidate the full potential of these analogs as novel Hh pathway inhibitors.

Supplementary Materials: The following supporting information can be downloaded at: https://www.mdpi.com/article/10.3390/molecules29133095/s1, Figure S1: Activity of Hh pathway agonist and antagonist controls in Hh-responsive cell lines C3H10T1/2 and *Gli*-reporter NIH3T3 cellular assays.; Figure S2: Screening of GANT61 analog set in C3H10T1/2 cells using AP absorbance assay.; Figure S3: Effects of selected GANT61 analogs in C3H10T1/2 stimulated by different Smo agonists.; Figure S4: Time course of *Gli1* mRNA expression in C3H10T1/2 cells.; Figure S5: Quantification of *GLI1* and *GLI2* mRNA expression in tumor cells.; Figure S6: Molecular docking prediction for the binding mode of GANT61 to GLI1.; Table S1: Compound ID and chemical structures for the GANT61 analog compound set used in this study.; Table S2: Top 12 predicted poses for unbiased docking of BAS 07019774 to GLI1-ZF.; Table S3: Table of Top 15 predicted poses for biased docking of BAS 07019774 to GLI1-ZF at Glu119/Glu167.

Author Contributions: Conceptualization, K.P.W. and M.S.D.; methodology, D.A.R., L.C., D.R.L. and M.S.D.; software, W.Z.; formal analysis, D.A.R., C.P.L. and W.Z.; investigation, D.A.R., L.C., D.R.L., M.T., N.E., G.R.S. and M.S.D.; writing—original draft preparation, K.P.W. and D.A.R.; writing—review and editing, D.A.R., C.P.L., M.S.D. and K.P.W.; visualization, W.Z.; supervision, K.P.W.; funding acquisition, K.P.W. All authors have read and agreed to the published version of the manuscript.

Funding: This research was funded by the National Institutes of Health, grant numbers U54AA030451, R01MD017405, and RCMI U54MD012392. Additional funding was obtained from the Golden LEAF Foundation and the BIOIMPACT Initiative of the State of North Carolina through the Biomanufacturing Research Institute & Technology Enterprise (BRITE) Center for Excellence at North Carolina Central University.

Institutional Review Board Statement: Not applicable.

Informed Consent Statement: Not applicable.

Data Availability Statement: The raw data supporting the conclusions of this article will be made available by the authors upon request.

Acknowledgments: Thanks to Denis Machado (UNC Charlotte) and Danni Harris (RTI) for their discussions on our docking studies. The analog similarity searching schematic in Figure 1 and graphical abstract were created with BioRender.com.

Conflicts of Interest: The authors declare no conflicts of interest.

References

1. Kinzler, K.W.; Bigner, S.H.; Bigner, D.D.; Trent, J.M.; Law, M.L.; O'Brien, S.J.; Wong, A.J.; Vogelstein, B. Identification of an amplified, highly expressed gene in a human glioma. *Science* **1987**, *236*, 70–73. [CrossRef]
2. Kinzler, K.W.; Ruppert, J.M.; Bigner, S.H.; Vogelstein, B. The GLI gene is a member of the Kruppel family of zinc finger proteins. *Nature* **1988**, *332*, 371–374. [CrossRef]
3. Kinzler, K.W.; Vogelstein, B. The GLI gene encodes a nuclear protein which binds specific sequences in the human genome. *Mol. Cell. Biol.* **1990**, *10*, 634–642. [CrossRef]
4. Kasper, M.; Regl, G.; Frischauf, A.M.; Aberger, F. GLI transcription factors: Mediators of oncogenic Hedgehog signalling. *Eur. J. Cancer* **2006**, *42*, 437–445. [CrossRef]
5. Ruppert, J.M.; Kinzler, K.W.; Wong, A.J.; Bigner, S.H.; Kao, F.T.; Law, M.L.; Seuanez, H.N.; O'Brien, S.J.; Vogelstein, B. The GLI-Kruppel family of human genes. *Mol. Cell. Biol.* **1988**, *8*, 3104–3113. [CrossRef]
6. Ingham, P.W. Hedgehog signaling. *Curr. Top. Dev. Biol.* **2022**, *149*, 1–58. [PubMed]
7. McMahon, A.P.; Ingham, P.W.; Tabin, C.J. Developmental roles and clinical significance of hedgehog signaling. *Curr. Top. Dev. Biol.* **2003**, *53*, 1–114. [CrossRef]
8. Zhang, Y.; Beachy, P.A. Cellular and molecular mechanisms of Hedgehog signalling. *Nat. Rev. Mol. Cell Biol.* **2023**, *24*, 668–687. [CrossRef] [PubMed]
9. Jing, J.; Wu, Z.; Wang, J.; Luo, G.; Lin, H.; Fan, Y.; Zhou, C. Hedgehog signaling in tissue homeostasis, cancers, and targeted therapies. *Signal Trans. Targeted Ther.* **2023**, *8*, 315. [CrossRef] [PubMed]
10. Rubin, L.L.; de Sauvage, F.J. Targeting the Hedgehog pathway in cancer. *Nat. Rev. Drug Discov.* **2006**, *5*, 1026–1033. [CrossRef]
11. Sigafoos, A.N.; Paradise, B.D.; Fernandez-Zapico, M.E. Hedgehog/GLI signaling pathway: Transduction, regulation, and implications for disease. *Cancers* **2021**, *13*, 3410. [CrossRef] [PubMed]

12. Gu, D.; Xie, J. Non-canonical Hh signaling in cancer—Current understanding and future directions. *Cancers* **2015**, *7*, 1684–1698. [CrossRef] [PubMed]
13. Jenkins, D. Hedgehog signalling: Emerging evidence for non-canonical pathways. *Cell. Signal.* **2009**, *21*, 1023–1034. [CrossRef] [PubMed]
14. Lauth, M.; Toftgård, R. Non-canonical activation of GLI transcription factors: Implications for targeted anti-cancer therapy. *Cell Cycle* **2007**, *6*, 2458–2463. [CrossRef]
15. Pietrobono, S.; Gagliardi, S.; Stecca, B. Non-canonical hedgehog signaling pathway in cancer: Activation of GLI transcription factors beyond smoothened. *Front. Genet.* **2019**, *10*, 556. [CrossRef] [PubMed]
16. Suchors, C.; Kim, J. Canonical hedgehog pathway and noncanonical GLI transcription factor activation in cancer. *Cells* **2022**, *11*, 2523. [CrossRef] [PubMed]
17. Niewiadomski, P.; Niedziółka, S.M.; Markiewicz, Ł.; Uśpieński, T.; Baran, B.; Chojnowska, K. Gli proteins: Regulation in development and cancer. *Cells* **2019**, *8*, 147. [CrossRef]
18. Zhu, H.; Lo, H.-W. The human glioma-associated oncogene homolog 1 (GLI1) family of transcription factors in gene regulation and diseases. *Curr. Genom.* **2010**, *11*, 238–245. [CrossRef]
19. Wu, J.; Di, D.; Zhao, C.; Liu, Y.; Chen, H.; Gong, Y.; Zhao, X.; Chen, H. Role of glioma-associated GLI1 oncogene in carcinogenesis and cancer-targeted therapy. *Curr. Cancer Drug Targets* **2018**, *18*, 558–566. [CrossRef]
20. Avery, J.T.; Zhang, R.; Boohaker, R.J. GLI1: A therapeutic target for cancer. *Front. Oncol.* **2021**, *11*, 673154. [CrossRef]
21. Miller, J.S.; Rhoades, J. Targeting hedgehog-driven mechanisms of drug-resistant cancers. *Front. Mol. Biosci.* **2023**, *10*, 1286090. [CrossRef] [PubMed]
22. Dennler, S.; Andre, J.; Alexaki, I.; Li, A.; Magnaldo, T.; ten Dijke, P.; Wang, X.J.; Verrecchia, F.; Mauviel, A. Induction of sonic hedgehog mediators by transforming growth factor-beta: Smad3-dependent activation of Gli2 and Gli1 expression in vitro and in vivo. *Cancer Res.* **2007**, *67*, 6981–6986. [CrossRef]
23. Wang, Y.; Ding, Q.; Yen, C.-J.; Xia, W.; Izzo, J.G.; Lang, J.-Y.; Li, C.-W.; Hsu, J.L.; Miller, S.A.; Wang, X. The crosstalk of mTOR/S6K1 and Hedgehog pathways. *Cancer Cell* **2012**, *21*, 374–387. [CrossRef]
24. Gruber Filbin, M.; Dabral, S.K.; Pazyra-Murphy, M.F.; Ramkissoon, S.; Kung, A.L.; Pak, E.; Chung, J.; Theisen, M.A.; Sun, Y.; Franchetti, Y. Coordinate activation of Shh and PI3K signaling in PTEN-deficient glioblastoma: New therapeutic opportunities. *Nat. Med.* **2013**, *19*, 1518–1523. [CrossRef]
25. Kasiri, S.; Shao, C.; Chen, B.; Wilson, A.N.; Yenerall, P.; Timmons, B.C.; Girard, L.; Tian, H.; Behrens, C.; Wistuba, I.I. GLI1 blockade potentiates the antitumor activity of PI3K antagonists in lung squamous cell carcinoma. *Cancer Res.* **2017**, *77*, 4448–4459. [CrossRef] [PubMed]
26. Chahal, K.K.; Parle, M.; Abagyan, R. Hedgehog pathway and smoothened inhibitors in cancer therapies. *Anti-Cancer Drugs* **2018**, *29*, 387–401. [CrossRef]
27. Nicheperovich, A.; Townsend-Nicholson, A. Towards Precision Oncology: The Role of Smoothened and Its Variants in Cancer. *J. Pers. Med.* **2022**, *12*, 1648. [CrossRef]
28. Burness, C.B. Sonidegib: First global approval. *Drugs* **2015**, *75*, 1559–1566. [CrossRef]
29. Hoy, S.M. Glasdegib: First global approval. *Drugs* **2019**, *79*, 207–213. [CrossRef] [PubMed]
30. Sheridan, C. Genentech obtains proof of concept for hedgehog inhibition. *Nat. Biotechnol.* **2009**, *27*, 968–970. [CrossRef] [PubMed]
31. Atwood, S.X.; Sarin, K.Y.; Whitson, R.J.; Li, J.R.; Kim, G.; Rezaee, M.; Ally, M.S.; Kim, J.; Yao, C.; Chang, A.L.; et al. Smoothened variants explain the majority of drug resistance in basal cell carcinoma. *Cancer Cell* **2015**, *27*, 342–353. [CrossRef]
32. Di Magno, L.; Coni, S.; Di Marcotullio, L.; Canettieri, G. Digging a hole under Hedgehog: Downstream inhibition as an emerging anticancer strategy. *Biochim. Biophys. Acta-Rev. Cancer* **2015**, *1856*, 62–72. [CrossRef] [PubMed]
33. Gonnissen, A.; Isebaert, S.; Haustermans, K. Targeting the Hedgehog signaling pathway in cancer: Beyond Smoothened. *Oncotarget* **2015**, *6*, 13899. [CrossRef]
34. Infante, P.; Alfonsi, R.; Botta, B.; Mori, M.; Di Marcotullio, L. Targeting GLI factors to inhibit the Hedgehog pathway. *Trends Pharmacol. Sci.* **2015**, *36*, 547–558. [CrossRef]
35. Sabol, M.; Trnski, D.; Musani, V.; Ozretić, P.; Levanat, S. Role of GLI transcription factors in pathogenesis and their potential as new therapeutic targets. *Intl. J. Mol. Sci.* **2018**, *19*, 2562. [CrossRef]
36. Yang, F.; Wynn, D.T.; Shen, C.; Ayad, N.G.; Robbins, D.J. Multiprotein GLI Transcriptional Complexes as Therapeutic Targets in Cancer. *Life* **2022**, *12*, 1967. [CrossRef]
37. Didiasova, M.; Schaefer, L.; Wygrecka, M. Targeting GLI transcription factors in cancer. *Molecules* **2018**, *23*, 1003. [CrossRef] [PubMed]
38. Ruiz i Altaba, A.; Ruiz i Altaba, A. How the Hedgehog Outfoxed the Crab: Interference with HEDGEHOG-GLI Signaling as Anti-Cancer Therapy? In *Madame Curie Bioscience Database*; Landes Bioscience: Austin, TX, USA, 2006; pp. 1–22.
39. Infante, P.; Mori, M.; Alfonsi, R.; Ghirga, F.; Aiello, F.; Toscano, S.; Di Marcotullio, L. Gli1/DNA interaction is a druggable target for Hedgehog-dependent tumors. *EMBO J.* **2015**, *34*, 200–217. [CrossRef] [PubMed]
40. Lauth, M.; Bergstrom, A.; Shimokawa, T.; Toftgard, R. Inhibition of GLI-mediated transcription and tumor cell growth by small-molecule antagonists. *Proc. Natl. Acad. Sci. USA* **2007**, *104*, 8455–8460. [CrossRef]
41. Agyeman, A.; Jha, B.K.; Mazumdar, T.; Houghton, J.A. Mode and specificity of binding of the small molecule GANT61 to GLI determines inhibition of GLI-DNA binding. *Oncotarget* **2014**, *5*, 4492–4503. [CrossRef]

42. Mazumdar, T.; DeVecchio, J.; Shi, T.; Jones, J.; Agyeman, A.; Houghton, J.A. Hedgehog signaling drives cellular survival in human colon carcinoma cells. *Cancer Res.* **2011**, *71*, 1092–1102. [CrossRef] [PubMed]
43. Riaz, S.K.; Khan, J.S.; Shah, S.T.A.; Wang, F.; Ye, L.; Jiang, W.G.; Malik, M.F.A. Involvement of hedgehog pathway in early onset, aggressive molecular subtypes and metastatic potential of breast cancer. *Cell Commun. Signal.* **2018**, *16*, 3. [CrossRef] [PubMed]
44. Wickstrom, M.; Dyberg, C.; Shimokawa, T.; Milosevic, J.; Baryawno, N.; Fuskevag, O.M.; Larsson, R.; Kogner, P.; Zaphiropoulos, P.G.; Johnsen, J.I. Targeting the hedgehog signal transduction pathway at the level of GLI inhibits neuroblastoma cell growth in vitro and in vivo. *Int. J. Cancer* **2013**, *132*, 1516–1524. [CrossRef] [PubMed]
45. Calcaterra, A.; Iovine, V.; Botta, B.; Quaglio, D.; D'Acquarica, I.; Ciogli, A.; Iazzetti, A.; Alfonsi, R.; Lospinoso Severini, L.; Infante, P.; et al. Chemical, computational and functional insights into the chemical stability of the Hedgehog pathway inhibitor GANT61. *J. Enzym. Inhib. Med. Chem.* **2018**, *33*, 349–358. [CrossRef]
46. Quaglio, D.; Infante, P.; Di Marcotullio, L.; Botta, B.; Mori, M. Hedgehog signaling pathway inhibitors: An updated patent review (2015–present). *Expert Opin. Ther. Pat.* **2020**, *30*, 235–250. [CrossRef]
47. Peer, E.; Tesanovic, S.; Aberger, F. Next-generation Hedgehog/GLI pathway inhibitors for cancer therapy. *Cancers* **2019**, *11*, 538. [CrossRef]
48. Hyman, J.M.; Firestone, A.J.; Heine, V.M.; Zhao, Y.; Ocasio, C.A.; Han, K.; Sun, M.; Rack, P.G.; Sinha, S.; Wu, J.J.; et al. Small-molecule inhibitors reveal multiple strategies for Hedgehog pathway blockade. *Proc. Natl. Acad. Sci. USA* **2009**, *106*, 14132–14137. [CrossRef] [PubMed]
49. Beauchamp, E.M.; Ringer, L.; Bulut, G.; Sajwan, K.P.; Hall, M.D.; Lee, Y.C.; Peaceman, D.; Ozdemirli, M.; Rodriguez, O.; Macdonald, T.J.; et al. Arsenic trioxide inhibits human cancer cell growth and tumor development in mice by blocking Hedgehog/GLI pathway. *J. Clin. Investig.* **2011**, *121*, 148–160. [CrossRef]
50. Carpenter, R.L.; Ray, H. Safety and Tolerability of Sonic Hedgehog Pathway Inhibitors in Cancer. *Drug Saf.* **2019**, *42*, 263–279. [CrossRef]
51. Didiasova, M.; Singh, R.; Wilhelm, J.; Kwapiszewska, G.; Wujak, L.; Zakrzewicz, D.; Schaefer, L.; Markart, P.; Seeger, W.; Lauth, M. Pirfenidone exerts antifibrotic effects through inhibition of GLI transcription factors. *FASEB J.* **2017**, *31*, 1916–1928. [CrossRef] [PubMed]
52. Li, B.; Fei, D.L.; Flaveny, C.A.; Dahmane, N.; Baubet, V.; Wang, Z.; Bai, F.; Pei, X.-H.; Rodriguez-Blanco, J.; Hang, B. Pyrvinium attenuates Hedgehog signaling downstream of smoothened. *Cancer Res.* **2014**, *74*, 4811–4821. [CrossRef] [PubMed]
53. Mahindroo, N.; Connelly, M.C.; Punchihewa, C.; Kimura, H.; Smeltzer, M.P.; Wu, S.; Fujii, N. Structure−Activity Relationships and Cancer-Cell Selective Toxicity of Novel Inhibitors of Glioma-Associated Oncogene Homologue 1 (Gli1) Mediated Transcription. *J. Med. Chem.* **2009**, *52*, 4277–4287. [CrossRef] [PubMed]
54. Bosco-Clément, G.; Zhang, F.; Chen, Z.; Zhou, H.-M.; Li, H.; Mikami, I.; Hirata, T.; Yagui-Beltran, A.; Lui, N.; Do, H.T. Targeting Gli transcription activation by small molecule suppresses tumor growth. *Oncogene* **2014**, *33*, 2087–2097. [CrossRef] [PubMed]
55. Actis, M.; Connelly, M.C.; Mayasundari, A.; Punchihewa, C.; Fujii, N. A structure–activity relationship study of small-molecule inhibitors of GLI1-mediated transcription. *Biopolymers* **2011**, *95*, 24–30. [CrossRef] [PubMed]
56. Mahindroo, N.; Connelly, M.C.; Punchihewa, C.; Yang, L.; Yan, B.; Fujii, N. Amide conjugates of ketoprofen and indole as inhibitors of Gli1-mediated transcription in the Hedgehog pathway. *Bioorg. Med. Chem.* **2010**, *18*, 4801–4811. [CrossRef] [PubMed]
57. Nayak, A.; Satapathy, S.R.; Das, D.; Siddharth, S.; Tripathi, N.; Bharatam, P.V.; Kundu, C. Nanoquinacrine induced apoptosis in cervical cancer stem cells through the inhibition of hedgehog-GLI1 cascade: Role of GLI-1. *Sci. Rep.* **2016**, *6*, 20600. [CrossRef]
58. Hosoya, T.; Arai, M.A.; Koyano, T.; Kowithayakorn, T.; Ishibashi, M. Naturally occurring small-molecule inhibitors of hedgehog/GLI-mediated transcription. *ChemBioChem* **2008**, *9*, 1082–1092. [CrossRef] [PubMed]
59. Arai, M.A.; Ochi, F.; Makita, Y.; Chiba, T.; Higashi, K.; Suganami, A.; Tamura, Y.; Toida, T.; Iwama, A.; Sadhu, S.K.; et al. GLI1 Inhibitors Identified by Target Protein Oriented Natural Products Isolation (TPO-NAPI) with Hedgehog Inhibition. *ACS Chem. Biol.* **2018**, *13*, 2551–2559. [CrossRef]
60. Li, X.Y.; Zhou, L.F.; Gao, L.J.; Wei, Y.; Xu, S.F.; Chen, F.Y.; Huang, W.J.; Tan, W.F.; Ye, Y.P. Cynanbungeigenin C and D, a pair of novel epimers from *Cynanchum bungei*, suppress hedgehog pathway-dependent medulloblastoma by blocking signaling at the level of Gli. *Cancer Lett.* **2018**, *420*, 195–207. [CrossRef]
61. Kim, J.; Lee, J.J.; Kim, J.; Gardner, D.; Beachy, P.A. Arsenic antagonizes the Hedgehog pathway by preventing ciliary accumulation and reducing stability of the Gli2 transcriptional effector. *Proc. Natl. Acad. Sci. USA* **2010**, *107*, 13432–13437. [CrossRef]
62. Wang, L.C.; Liu, Z.Y.; Gambardella, L.; Delacour, A.; Shapiro, R.; Yang, J.; Sizing, I.; Rayhorn, P.; Garber, E.A.; Benjamin, C.D.; et al. Regular articles: Conditional disruption of hedgehog signaling pathway defines its critical role in hair development and regeneration. *J. Investig. Dermatol.* **2000**, *114*, 901–908. [CrossRef] [PubMed]
63. Williams, K.P.; Rayhorn, P.; Chi-Rosso, G.; Garber, E.A.; Strauch, K.L.; Horan, G.S.; Reilly, J.O.; Baker, D.P.; Taylor, F.R.; Koteliansky, V.; et al. Functional antagonists of sonic hedgehog reveal the importance of the N terminus for activity. *J. Cell Sci.* **1999**, *112 Pt 23*, 4405–4414. [CrossRef] [PubMed]
64. Reznikoff, C.A.; Brankow, D.W.; Heidelberger, C. Establishment and characterization of a cloned line of C3H mouse embryo cells sensitive to postconfluence inhibition of division. *Cancer Res.* **1973**, *33*, 3231–3238. [PubMed]
65. Nakamura, T.; Aikawa, T.; Iwamoto-Enomoto, M.; Iwamoto, M.; Higuchi, Y.; Maurizio, P.; Kinto, N.; Yamaguchi, A.; Noji, S.; Kurisu, K. Induction of Osteogenic Differentiation by Hedgehog Proteins. *Biochem. Biophys. Res. Comm.* **1997**, *237*, 465–469. [CrossRef] [PubMed]

66. Williams, K.; Scott, J. Enzyme assay design for high-throughput screening. *Meth. Mol. Biol.* **2009**, *565*, 107–126.
67. Sinha, S.; Chen, J.K. Purmorphamine activates the Hedgehog pathway by targeting Smoothened. *Nat. Chem. Biol.* **2006**, *2*, 29–30. [CrossRef]
68. Wu, X.; Walker, J.; Zhang, J.; Ding, S.; Schultz, P.G. Purmorphamine induces osteogenesis by activation of the hedgehog signaling pathway. *Chem. Biol.* **2004**, *11*, 1229–1238. [CrossRef]
69. Taipale, J.; Chen, J.K.; Cooper, M.K.; Wang, B.; Mann, R.K.; Milenkovic, L.; Scott, M.P.; Beachy, P.A. Effects of oncogenic mutations in Smoothened and Patched can be reversed by cyclopamine. *Nature* **2000**, *406*, 1005–1009. [CrossRef] [PubMed]
70. Abraham, V.C.; Towne, D.L.; Waring, J.F.; Warrior, U.; Burns, D.J. Application of a high-content multiparameter cytotoxicity assay to prioritize compounds based on toxicity potential in humans. *SLAS Discov.* **2008**, *13*, 527–537. [CrossRef]
71. Li, J.; Cai, J.; Zhao, S.; Yao, K.; Sun, Y.; Li, Y.; Chen, L.; Li, R.; Zhai, X.; Zhang, J. GANT61, a GLI inhibitor, sensitizes glioma cells to the temozolomide treatment. *J. Exp. Clin. Cancer Res.* **2016**, *35*, 184. [CrossRef]
72. Melamed, J.R.; Morgan, J.T.; Ioele, S.A.; Gleghorn, J.P.; Sims-Mourtada, J.; Day, E.S. Investigating the role of Hedgehog/GLI1 signaling in glioblastoma cell response to temozolomide. *Oncotarget* **2018**, *9*, 27000. [CrossRef] [PubMed]
73. Volnitskiy, A.; Shtam, T.; Burdakov, V.; Kovalev, R.; Konev, A.; Filatov, M. Abnormal activity of transcription factors gli in high-grade gliomas. *PLoS ONE* **2019**, *14*, e0211980. [CrossRef] [PubMed]
74. Huang, L.; Walter, V.; Hayes, D.N.; Onaitis, M. Hedgehog–GLI signaling inhibition suppresses tumor growth in squamous lung cancer. *Clin. Cancer Res.* **2014**, *20*, 1566–1575. [CrossRef] [PubMed]
75. Pavletich, N.P.; Pabo, C.O. Crystal structure of a five-finger GLI-DNA complex: New perspectives on zinc fingers. *Science* **1993**, *261*, 1701–1707. [CrossRef] [PubMed]
76. Steiner, T.; Koellner, G. Hydrogen bonds with π-acceptors in proteins: Frequencies and role in stabilizing local 3D structures. *J. Mol. Biol.* **2001**, *305*, 535–557. [CrossRef]
77. Benvenuto, M.; Masuelli, L.; De Smaele, E.; Fantini, M.; Mattera, R.; Cucchi, D.; Bonanno, E.; Di Stefano, E.; Frajese, G.V.; Orlandi, A.; et al. In vitro and in vivo inhibition of breast cancer cell growth by targeting the Hedgehog/GLI pathway with SMO (GDC-0449) or GLI (GANT-61) inhibitors. *Oncotarget* **2016**, *7*, 9250–9270. [CrossRef] [PubMed]
78. Desch, P.; Asslaber, D.; Schnidar, H.; Mangelberger, D.; Alinger, B.; Stoecher, M.; Hofbauer, S.W.; Neureiter, D.; Tinhofer, I. Inhibition of GLI, but not Smoothened, induces apoptosis in chronic lymphocytic leukemia cells. *Oncogene* **2010**, *29*, 4885. [CrossRef] [PubMed]
79. Mazumdar, T.; Devecchio, J.; Agyeman, A.; Shi, T.; Houghton, J.A. Blocking Hedgehog survival signaling at the level of the GLI genes induces DNA damage and extensive cell death in human colon carcinoma cells. *Cancer Res.* **2011**, *71*, 5904–5914. [CrossRef]
80. Curran, T. Reproducibility of academic preclinical translational research: Lessons from the development of Hedgehog pathway inhibitors to treat cancer. *Open Biol.* **2018**, *8*, 180098. [CrossRef]
81. Bacelar Sacramento de Araújo, T.; de Oliveira Siquara da Rocha, L.; Torres Andion Vidal, M.; Cerqueira Coelho, P.L.; Galvão dos Reis, M.; Solano de Freitas Souza, B.; Botelho Pereira Soares, M.; Almeida Pereira, T.; Della Coletta, R.; Pereira Bezerra, D. GANT61 Reduces Hedgehog Molecule (GLI1) Expression and Promotes Apoptosis in Metastatic Oral Squamous Cell Carcinoma Cells. *Int. J. Mol. Sci.* **2020**, *21*, 6076. [CrossRef]
82. Dash, R.C.; Wen, J.; Zaino, A.M.; Morel, S.R.; Chau, L.Q.; Wechsler-Reya, R.J.; Hadden, M.K. Structure-based virtual screening identifies an 8-hydroxyquinoline as a small molecule GLI1 inhibitor. *Mol. Ther. Oncolytics* **2021**, *20*, 265–276. [CrossRef]
83. Li, X.; Sun, R.; Wu, H.; Zheng, C.; Long, Y.-Q. Targeting the Hedgehog pathway with novel Gli1 hydrophobic tagging degraders. *Bioorg. Chem.* **2023**, *138*, 106649. [CrossRef]
84. Benet, L.Z.; Hosey, C.M.; Ursu, O.; Oprea, T.I. BDDCS, the Rule of 5 and drugability. *Adv. Drug Deliv. Rev.* **2016**, *101*, 89–98. [CrossRef] [PubMed]
85. Lipinski, C.A.; Lombardo, F.; Dominy, B.W.; Feeney, P.J. Experimental and computational approaches to estimate solubility and permeability in drug discovery and development settings. *Adv. Drug Deliv. Rev.* **1997**, *23*, 3–25. [CrossRef]
86. Berardozzi, S.; Bernardi, F.; Infante, P.; Ingallina, C.; Toscano, S.; De Paolis, E.; Alfonsi, R.; Caimano, M.; Botta, B.; Mori, M.; et al. Synergistic inhibition of the Hedgehog pathway by newly designed Smo and Gli antagonists bearing the isoflavone scaffold. *Eur. J. Med. Chem.* **2018**, *156*, 554–562. [CrossRef] [PubMed]
87. Falsini, A.; Giuntini, G.; Mori, M.; Ghirga, F.; Quaglio, D.; Cucinotta, A.; Coppola, F.; Filippi, I.; Naldini, A.; Botta, B. Hedgehog Pathway Inhibition by Novel Small Molecules Impairs Melanoma Cell Migration and Invasion under Hypoxia. *Pharmaceuticals* **2024**, *17*, 227. [CrossRef]
88. Lospinoso Severini, L.; Ghirga, F.; Bufalieri, F.; Quaglio, D.; Infante, P.; Di Marcotullio, L. The SHH/GLI signaling pathway: A therapeutic target for medulloblastoma. *Expert Opin. Ther. Targets* **2020**, *24*, 1159–1181. [CrossRef]
89. Zhang, R.; Ma, J.; Avery, J.T.; Sambandam, V.; Nguyen, T.H.; Xu, B.; Suto, M.J.; Boohaker, R.J. GLI1 Inhibitor SRI-38832 attenuates chemotherapeutic resistance by downregulating NBS1 transcription in BRAFV600E colorectal cancer. *Front. Oncol.* **2020**, *10*, 241. [CrossRef]
90. Manetti, F.; Stecca, B.; Santini, R.; Maresca, L.; Giannini, G.; Taddei, M.; Petricci, E. Pharmacophore-Based Virtual Screening for Identification of Negative Modulators of GLI1 as Potential Anticancer Agents. *Med. Chem. Lett.* **2020**, *11*, 832–838. [CrossRef]
91. Manetti, F.; Maresca, L.; Crivaro, E.; Pepe, S.; Cini, E.; Singh, S.; Governa, P.; Maramai, S.; Giannini, G.; Stecca, B. Quinolines and Oxazino-quinoline Derivatives as Small Molecule GLI1 Inhibitors Identified by Virtual Screening. *ACS Med. Chem. Lett.* **2022**, *13*, 1329–1336. [CrossRef]

92. Maresca, L.; Crivaro, E.; Migliorini, F.; Anichini, G.; Giammona, A.; Pepe, S.; Poggialini, F.; Vagaggini, C.; Giannini, G.; Sestini, S. Targeting GLI1 and GLI2 with small molecule inhibitors to suppress GLI-dependent transcription and tumor growth. *Pharmacol. Res.* **2023**, *195*, 106858. [CrossRef] [PubMed]
93. Noser, A.A.; El-Barbary, A.; Salem, M.M.; El Salam, H.A.A.; Shahien, M. Synthesis and molecular docking simulations of novel azepines based on quinazolinone moiety as prospective antimicrobial and antitumor hedgehog signaling inhibitors. *Sci. Rep.* **2024**, *14*, 3530. [CrossRef] [PubMed]
94. Chen, J.K.; Taipale, J.; Young, K.E.; Maiti, T.; Beachy, P.A. Small molecule modulation of Smoothened activity. *Proc. Natl. Acad. Sci. USA* **2002**, *99*, 14071–14076. [CrossRef]
95. Frank-Kamenetsky, M.; Zhang, X.M.; Bottega, S.; Guicherit, O.; Wichterle, H.; Dudek, H.; Bumcrot, D.; Wang, F.Y.; Jones, S.; Shulok, J. Small-molecule modulators of Hedgehog signaling: Identification and characterization of Smoothened agonists and antagonists. *J. Biol.* **2002**, *1*, 10. [CrossRef] [PubMed]
96. Niyomrattanakit, P.; Abas, S.N.; Lim, C.C.; Beer, D.; Shi, P.-Y.; Chen, Y.-L. A Fluorescence-Based Alkaline Phosphatase–Coupled Polymerase Assay for Identification of Inhibitors of Dengue Virus RNA-Dependent RNA Polymerase. *J. Biomol. Screen.* **2011**, *16*, 201–210. [CrossRef] [PubMed]
97. House, A.J.; Daye, L.R.; Tarpley, M.; Addo, K.; Lamson, D.S.; Parker, M.K.; Williams, K.P. Design and characterization of a photo-activatable hedgehog probe that mimics the natural lipidated form. *Arch. Biochem. Biophys.* **2015**, *567*, 66–74. [CrossRef]
98. Tarpley, M.; Oladapo, H.O.; Strepay, D.; Caligan, T.B.; Chdid, L.; Shehata, H.; Roques, J.R.; Thomas, R.; Laudeman, C.P.; Onyenwoke, R.U. Identification of harmine and β-carboline analogs from a high-throughput screen of an approved drug collection; profiling as differential inhibitors of DYRK1A and monoamine oxidase A and for in vitro and in vivo anti-cancer studies. *Eur. J. Pharm. Sci.* **2021**, *162*, 105821. [CrossRef]

molecules

MDPI

Article

Sterol Derivatives Specifically Increase Anti-Inflammatory Oxylipin Formation in M2-like Macrophages by LXR-Mediated Induction of 15-LOX

Reiichi Ohno †, Malwina Mainka †, Rebecca Kirchhoff, Nicole M. Hartung and Nils Helge Schebb *

Chair of Food Chemistry, Faculty of Mathematics and Natural Sciences, University of Wuppertal, Gaußstr. 20, 42119 Wuppertal, Germany

* Correspondence: nils@schebb-web.de; Tel.: +49-202-439-3457

† These authors contributed equally to this work.

Abstract: The understanding of the role of LXR in the regulation of macrophages during inflammation is emerging. Here, we show that LXR agonist T09 specifically increases 15-LOX abundance in primary human M2 macrophages. In time- and dose-dependent incubations with T09, an increase of 3-fold for ALOX15 and up to 15-fold for 15-LOX-derived oxylipins was observed. In addition, LXR activation has no or moderate effects on the abundance of macrophage marker proteins such as TLR2, TLR4, PPARγ, and IL-1RII, as well as surface markers (CD14, CD86, and CD163). Stimulation of M2-like macrophages with FXR and RXR agonists leads to moderate ALOX15 induction, probably due to side activity on LXR. Finally, desmosterol, 24(*S*),25-Ep cholesterol and 22(*R*)-OH cholesterol were identified as potent endogenous LXR ligands leading to an ALOX15 induction. LXR-mediated ALOX15 regulation is a new link between the two lipid mediator classes sterols, and oxylipins, possibly being an important tool in inflammatory regulation through anti-inflammatory oxylipins.

Keywords: macrophages; liver X receptor; lipoxygenase; oxylipins; eicosanoids; specialized pro-resolving mediators

1. Introduction

Inflammation is a protective mechanism against infection or tissue injury. During the course of inflammation, immune cells, such as neutrophils and monocytes, are recruited in several phases and stimulate the formation and release of different cytokines, chemokines, growth factors, and lipid mediators [1–3]. In the inflammatory process, monocytes differentiate into macrophages, which have three main tasks: the production of immunomodulators, phagocytosis, and antigen presentation [4,5].

The phagocytosis of cell debris and apoptotic neutrophils as a consequence of acute inflammation serves to restore tissue homeostasis [5,6]. The produced and released immunomodulators, such as interleukin (IL)-1, IL-6, tumor necrosis factor (TNF)-α, interferon (IFN)-α/β, IL-10, IL-12, or IL-18, regulate immune responses, such as stimulation of proliferation of activated natural killer cells or regulation of leukocyte migration from blood to tissue [5,7,8]. In addition, T-cells are attracted by antigen presentation and the release of chemoattractants [5,6].

Primary cell culture of monocyte-derived macrophages allows us to investigate the signaling and regulatory pathways of the innate immune response. Under cell culture conditions, macrophage differentiation is mimicked using different stimuli to investigate the regulation of the arachidonic acid cascade. The (simplified) classification of macrophages is characterized by the expression pattern of characteristic markers, including surface markers as well as pro/anti-inflammatory proteins [9]. Stimulation with granulocyte/macrophage colony-stimulating factor (GM-CSF) leads to the development of a pro-inflammatory state with increased TNF expression and release. After priming with macrophage colony-stimulating factor (M-CSF), the cells exert an anti-inflammatory character with increased IL-10 expression and release. In addition, the polarization of macrophages is induced by stimuli such as cytokines or ligands of the innate immune response, such as bacterial lipopolysaccharide (LPS) [10]. Commonly, stimulation of macrophages with IFNγ and LPS is used to generate classically activated macrophages, i.e., M1-like macrophages [9,11]. The polarization of alternatively activated macrophages is induced by IL-4, IL-10, or IL-13, i.e., M2-like macrophages [9,12].

During the inflammatory process, the polarization of macrophages can switch from pro- to anti-inflammatory depending on the inflammatory environment [13–15]. A switch of the macrophage phenotype can be initiated by efferocytosis, i.e., the uptake of apoptotic cells [14], which is associated with nuclear receptors such as retinoid X receptors (RXR), PPAR, or liver X receptors (LXR) [16]. Moreover, PPAR is known to be involved in macrophage polarization and inflammatory processes [17]. While inactivation of PPAR or RXR impairs efferocytosis of apoptotic cells, LXR-deficient macrophages are unable to clear apoptotic cells altogether [18,19]. In addition, LXR also drives efferocytosis indirectly by increasing its own expression in an autoregulatory manner [20].

LXR belongs to the family of nuclear receptors involved in the regulation of metabolic homeostasis and inflammation [21]. There are two LXR isoforms that are expressed differently in tissues, despite their high structural similarity (77%). While LXRα is mainly expressed in the liver, kidney, intestine, adipose tissue, or macrophages, LXRβ can be found in all human cells [22,23]. LXR target genes include sterol response element binding protein 1c (SREBP1c), apolipoprotein (apo) E, ATP-binding cassette transporters (ABC)A1, ABCG1, ABCG5, cytochrome P-450 7A1 (CYP7A1), and fatty acid synthase (FAS) [24,25]. As LXR is involved in both cholesterol and lipid metabolism, it could be an important link in the regulation of inflammation by impacting the formation of anti-inflammatory lipid mediators and thus driving the inflammation towards the resolution phase.

The resolution phase macrophages produce a variety of anti-inflammatory cytokines and chemokines. In addition, the formation of lipid mediators—oxylipins—plays a decisive role in the course of inflammation. High levels of cyclooxygenase (COX)- and 5-lipoxygenase (LOX)-derived products from arachidonic acid (ARA) are well investigated pro-inflammatory oxylipins, e.g., prostaglandin (PG) E2 and D2 or leukotriene (LT) B4, which are predominantly formed and released at the beginning of the inflammation. In contrast, multi-hydroxylated metabolites of different polyunsaturated fatty acids (PUFA) such as ARA, eicosapentaenoic acid (EPA), and docosahexaenoic acid (DHA) are described to be formed during the resolution phase and have anti-inflammatory properties [26,27]. The formation of multi-hydroxylated oxylipins such as 5,15-diHETE and 5,12-diHETE predominantly involves different lipoxygenases (LOX) interacting with each other (Figure 1) and 15-LOX being a key enzyme [28]. Although there is doubt about the formation, detectability, and receptors of multi-hydroxylated LOX products, so-called specialized pro-resolving mediators (SPM) and SPM bearing three hydroxy groups are not formed in macrophages [29], the anti-inflammatory effect of 15-LOX is undisputed [30].

Recently, we have shown that stimulation of macrophages with the synthetic LXR agonist T0901317 (T09) increases the expression of anti-inflammatory genes, such as ALOX15 (15-LOX gene) [31]. Here, we aim to investigate the LXR-induced influence on the ARA cascade and formation of oxylipins in macrophages, as well as different markers involved in differentiation and polarization. For this, human peripheral blood mononuclear cells (PBMC) left untreated or differentiated into M1- and M2-like macrophages were stimulated with T09, and the formation of oxylipins, protein levels, and presence of surface markers were compared to non-stimulated cells. The specificity of nuclear receptor activation was investigated using FXR agonists and RXR agonists. Finally, sterols were identified as endogenous LXR ligands, elevating 15-LOX abundance and upregulating oxylipin formation. In summary, with the sterol-mediated induction of oxylipins, we demonstrated a new link between these lipid mediator classes in the regulation of M2-like macrophages.

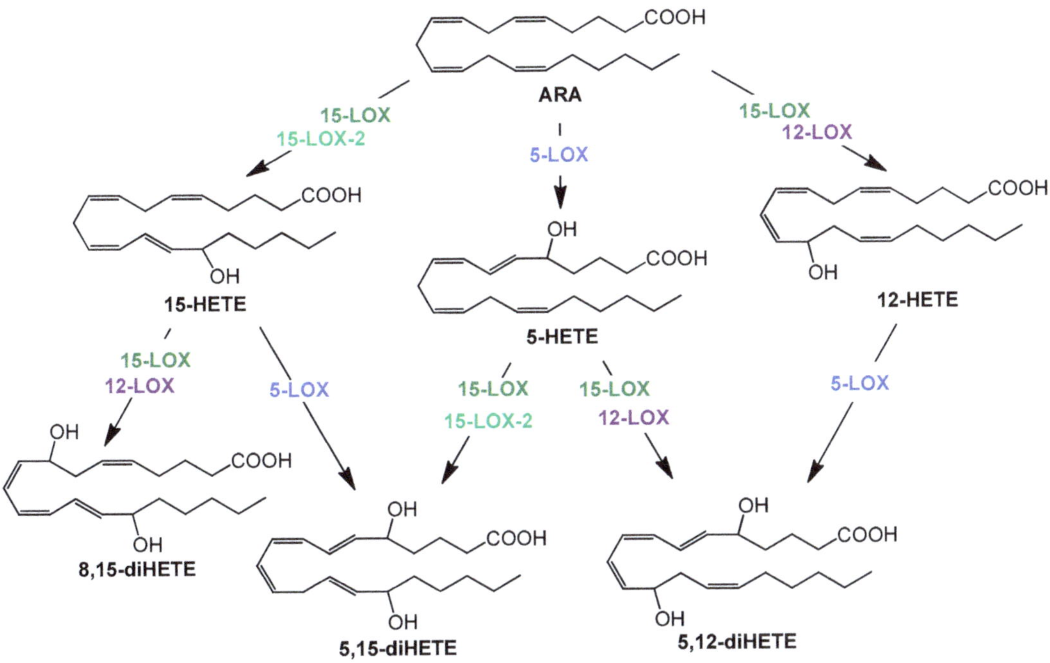

Figure 1. Simplified overview of structures and formation routes of multiple hydroxylated ARA metabolites by the human 5 (blue)-, 12 (purple)-, 15- lipoxygenase (dark green), and 15-LOX-2 (light green).

2. Results

In the present study, we investigated the effect of LXR activation by different nuclear receptor agonists and endogenous lipids on the induction of ARA cascade enzymes and oxylipin formation, as well as the overall effects on proteins and markers that are involved in the polarization of human primary macrophages.

PBMC were left untreated or differentiated into M1- or M2-like macrophages and stimulated with the synthetic LXR agonist T09 (1 µM) for 3 h (Figure 2A). Oxylipins were analyzed using a LC-MS/MS targeted oxylipin metabolomics method. Figure 2B shows representative oxylipins derived from ARA. The analysis of the protein levels was carried out by LC-MS/MS based targeted proteomics (Section 1 + Tables S1–S3, Supplementary Material). The effect of T09-mediated LXR activation on ARA cascade enzymes, receptors, and other protein levels specific for macrophages is depicted in Figure 2C. The presence of macrophage surface markers reflecting their polarization was analyzed by immunofluorescence staining (Figure 2D). T09 had no effect on the morphology or confluency of untreated M1- and M2-like macrophages.

The T09-induced 15-LOX abundance and activity were further evaluated in a time- and dose-dependent manner (Figure 3 and Table 1).

Following the evaluation of the optimal incubation conditions, the induction of 15-LOX by other LXR-, FXR-, and RXR-specific compounds as well as cholesterol derivatives (Figure 4) was investigated. The results are shown in Figure 5 and Table S4.

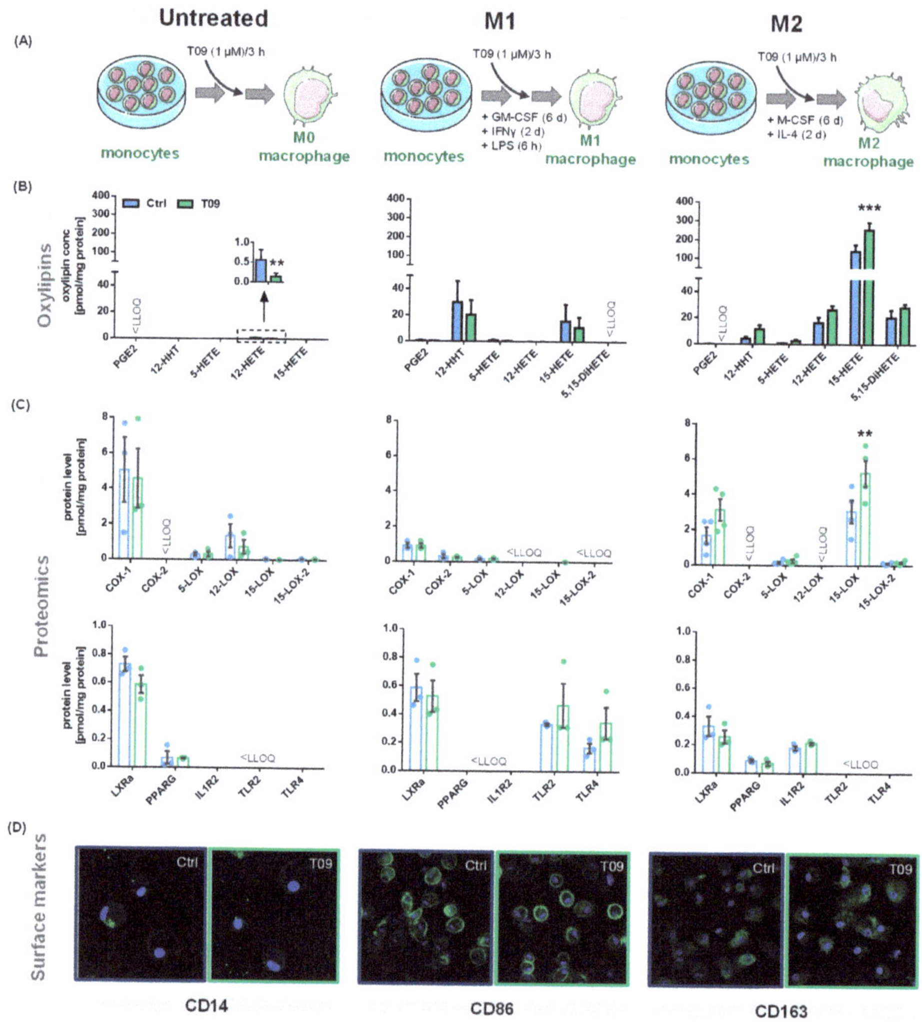

Figure 2. Effect of the LXR agonist T09 on the ARA cascade and differentiation of primary macrophages. (**A**) Primary human blood monocytic cells were treated with 10 ng/mL GM-CSF (M1 type) for 8 days or with M-CSF (M2 type) for 8 days, as well as 10 ng/mL IFNγ (M1 type) or IL-4 (M2 type) for the final 2 days. Additionally, M1 cells were challenged with 100 ng/mL LPS for 6 h. For the untreated macrophages, the adhered monocytes were left untreated for 8 days. The three different macrophage phenotypes were incubated with (blue) or without the synthetic LXR agonist T09 (1 μM; green) for the final three hours. Shown are (**B**) oxylipin concentrations and (**C**) protein levels (mean ± SEM; cells from 3–5 donors). (**D**) The three macrophage phenotypes with or without T09 (1 μM) were immunostained for specific macrophage markers CD14 (untreated), CD86 (M1), or CD163 (M2) using specific mouse anti-human antibodies (green) (magnification 63×). Nuclei were counterstained with Hoechst (blue). The immuno-positivity of the antibodies towards the respective macrophage types is shown in Figure S1. Differences were considered significant at p-values < 0.05 (**) or < 0.01 (***) using a two-tailed, unpaired student's t-test.

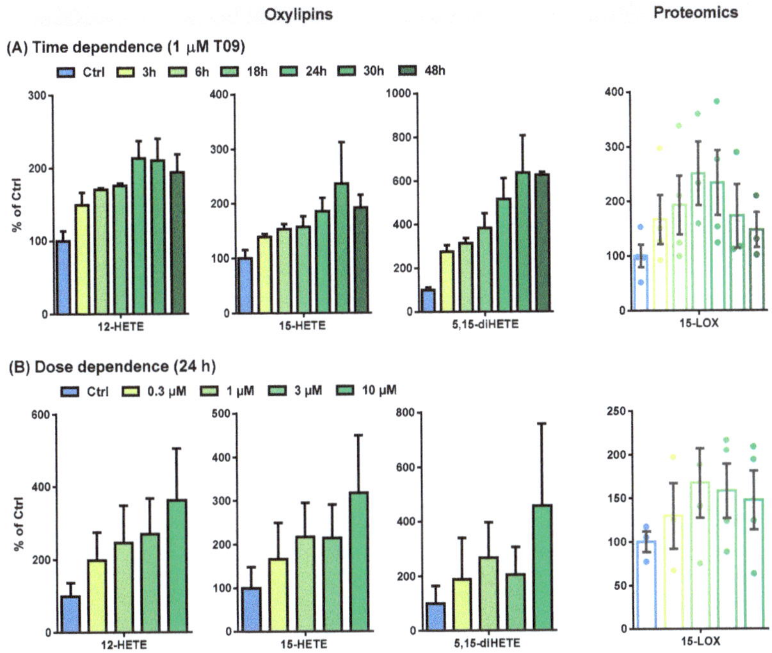

Figure 3. (**A**) Time- and (**B**) dose-dependent T09-induced 15-LOX abundance and activity in primary M2-like macrophages. Shown is the increase of ARA-derived 15-LOX metabolites 12-HETE, 15-HETE, and 5,15-diHETE (**left**) and 15-LOX abundance (**right**). Results are shown as % of Ctrl (mean ± SEM, cells from 3–5 donors).

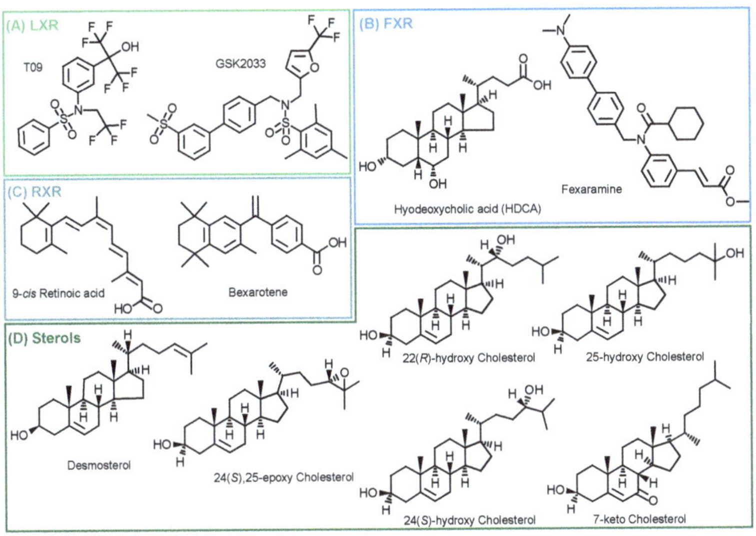

Figure 4. Structures of investigated compounds. (**A**) LXR agonist (T09) and antagonist (GSK2033); (**B**) FXR agonists (Fexaramine and Hyodeoxycholic acid); (**C**) RXR agonists (Bexarotene and 9-*cis* Retinoic acid); and (**D**) tested cholesterol precursor (desmosterol) and oxidized metabolites.

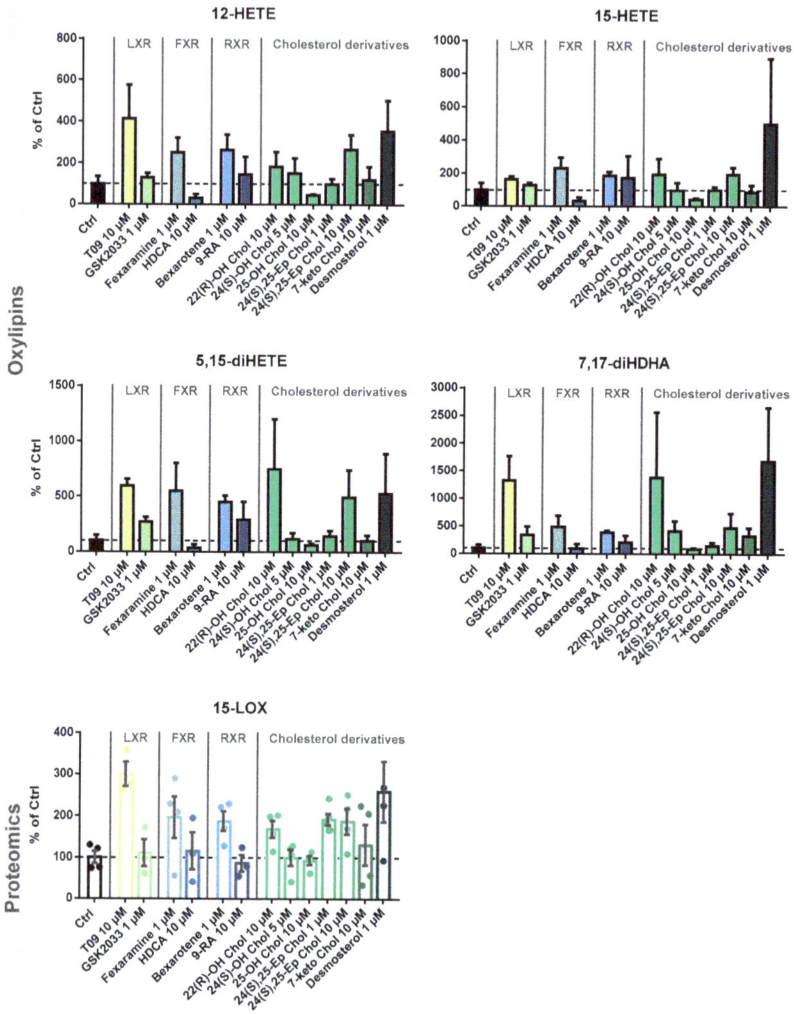

Figure 5. Investigation of 15-LOX induction by LXR, FXR, and RXR agonists and identification of cholesterol precursors and metabolites. M2-like macrophages were incubated with test compounds for 24 h. Shown is the increase of 15-LOX-derived mono-hydroxylated metabolites 12-HETE and 15-HETE (**top**) and multiple-hydroxylated metabolites 5,15-diHETE and 7,17-diHDHA (**middle**), as well as 15-LOX abundance (**bottom**). Results are shown as % of Ctrl (mean ± SEM, cells from 3–5 donors).

Table 1. Oxylipin concentrations and 15-LOX levels in M2-like macrophages. Primary blood monocytic cells were differentiated into M2-like macrophages with 10 ng/mL M-CSF for 8 days and incubated with IL-4 for the final 48 h. M2-like macrophages were incubated with 1 µM T09 for different periods of time to investigate the time-dependent correlation of 15-LOX abundance and activity (top) and with different T09 concentrations for 24 h to determine the dose-dependent correlation (bottom). The results are shown as mean ± SEM (n = 3–5).

		Time-Dependent Incubations [pmol/mg Protein]						
		Ctrl	3 h	6 h	18 h	24 h	30 h	48 h
Oxylipin concentration	12-HETE	15.5 ± 2.2	22.7 ± 2.2	26.4 ± 3.3	27.1 ± 3.4	32.5 ± 3.4	33.5 ± 8.9	29.3 ± 2.9
	15-HETE	184 ± 29	256 ± 40	277 ± 34	280 ± 25	327 ± 16	476 ± 213	364 ± 100
	5,15-diHETE	9.18 ± 1.1	25.6 ± 4.9	28.3 ± 2.8	33.7 ± 1.3	45.9 ± 8.2	59.6 ± 20	57.3 ± 6.4
	5,12-diHETE	0.47 ± 0.1	1.00 ± 0.4	1.23 ± 0.2	1.76 ± 0.4	2.17 ± 0.2	1.11 ± 0.4	0.64 ± 0.1
	5,15-diHEPE	3.30 ± 0.4	7.73 ± 0.4	9.60 ± 1.4	12.9 ± 1.9	14.0 ± 2.5	10.4 ± 0.9	10.2 ± 3.2
	8,15-diHETE	0.28 ± 0.1	0.55 ± 0.1	0.99 ± 0.3	0.98 ± 0.2	0.87 ± 0.2	0.83 ± 0.3	0.64 ± 0.5
	7,17-diHDHA	10.7 ± 1.0	21.1 ± 3.7	25.5 ± 0.8	33.1 ± 1.3	40.4 ± 1.7	56.9 ± 17	39.5 ± 5.3
	12-HEPE	1.60 ± 0.3	1.93 ± 0.2	2.29 ± 0.3	2.32 ± 0.1	2.11 ± 0.5	1.20 ± 0.9	1.23 ± 0.7
	15-HEPE	18.9 ± 2.9	22.1 ± 2.7	26.1 ± 3.1	28.83 ± 1.8	22.1 ± 4.1	11.7 ± 8.9	12.1 ± 8.5
	14-HDHA	16.2 ± 4.0	18.3 ± 3.3	22.3 ± 4.1	24.7 ± 2.0	25.0 ± 7.5	14.2 ± 12	14.4 ± 14
	7-HDHA	3.65 ± 1.6	2.46 ± 0.8	2.89 ± 1.0	2.86 ± 0.4	2.98 ± 1.1	2.13 ± 1.8	2.18 ± 1.5
	17-HDHA	45.4 ± 14	40.3 ± 8.5	51.4 ± 13	52.2 ± 5.9	50.7 ± 15	33.3 ± 29	34.0 ± 39
	15-LOX level	2.95 ± 0.6	4.40 ± 0.9	4.81 ± 0.6	5.66 ± 1.1	5.95 ± 0.8	2.04 ± 1.0	1.76 ± 0.6
		Dose-Dependent Incubations [pmol/mg Protein]						
		Ctrl	0.3 µM	1 µM	3 µM	10 µM		
Oxylipin concentration	12-HETE	2.78 ± 1.0	5.55 ± 2.1	6.86 ± 2.9	7.53 ± 2.7	10.1 ± 4.0		
	15-HETE	39.9 ± 20	66.2 ± 33	86.3 ± 31	85.3 ± 31	126 ± 53		
	5,15-diHETE	1.02 ± 0.7	1.93 ± 1.6	2.73 ± 1.3	2.10 ± 1.0	4.66 ± 3.1		
	5,12-diHETE	0.22 ± 0.1	0.20 ± 0.1	0.30 ± 0.1	0.23 ± 0.1	0.34 ± 0.2		
	5,15-diHEPE	0.37 ± 0.1	0.80 ± 0.2	1.11 ± 0.3	1.29 ± 0.5	1.68 ± 0.8		
	8,15-diHETE	0.14 ± 0.1	0.13 ± 0.1	0.20 ± 0.02	0.06 ± 0.02	0.20 ± 0.05		
	7,17-diHDHA	0.71 ± 0.4	1.79 ± 1.2	1.83 ± 0.9	2.34 ± 1.8	3.94 ± 2.6		
	12-HEPE	0.32 ± 0.14	0.80 ± 0.3	0.83 ± 0.3	0.80 ± 0.2	1.04 ± 0.3		
	15-HEPE	5.20 ± 2.7	9.29 ± 4.5	10.2 ± 3.8	9.00 ± 2.7	12.4 ± 4.4		
	14-HDHA	1.79 ± 0.8	4.51 ± 2.1	4.83 ± 2.1	5.43 ± 1.9	6.07 ± 2.2		
	7-HDHA	0.88 ± 0.5	1.63 ± 1.0	1.97 ± 0.9	1.60 ± 0.5	1.50 ± 0.6		
	17-HDHA	6.32 ± 3.4	14.4 ± 8.4	15.5 ± 6.2	15.4 ± 5.3	18.4 ± 7.1		
	15-LOX level	0.77 ± 0.1	1.00 ± 0.2	1.29 ± 0.2	1.22 ± 0.1	1.13 ± 0.1		

3. Discussion

The nuclear receptor LXR is associated with different immune cell functions [32]. We have recently shown that stimulation of M2-like macrophages with LXR agonist T09 leads to overexpression of ALOX15 as well as increased formation of 15-LOX-derived oxylipins [31]. 15-LOX as well as its products, such as 15-HETE or 15-HEPE, are discussed as having anti-inflammatory properties [30,33–35].

Here, we investigated the effect of LXR activation on the enzymes of the ARA cascade and oxylipin formation, as well as the effect on markers of human primary macrophages involved in polarization, in detail (Figure 2A). Our experiments demonstrate that in IL-4-stimulated M2-like macrophages, 15-LOX abundance is increased by T09 (Figure 2B,C). The LXR agonist T09 has a specific effect on 15-LOX abundance and its metabolites (15-HETE, 5,15-diHETE). T09 showed no or a low effect on PPARγ, IL-1RII (Figure 2C), and the surface marker CD163 (Figure 2D), which are highly expressed in M2-like macrophages [36,37], and thus this indicates that polarization of M2 macrophages was not influenced by T09-stimulation.

In M1-like macrophages, T09 also has a neglectable effect on the investigated markers involved in the differentiation process (Figure 2C,D). The surface marker CD86, which is highly expressed in M1-like macrophages after IFNγ stimulation [38], remained unchanged.

Stimulation with T09 had no effect on the 15-LOX-derived 15-HETE and did not increase ALOX15 abundance in these cells. Stimulation with T09 hardly modulated M1-specific proteins, with a trend towards higher TLR2 and TLR4 amounts and a decrease of COX-2 and its products PGE2 and 12-HHT (Figure 2B,C).

The untreated macrophages have low levels of PPARγ and the surface marker CD14 remained unchanged after stimulation with T09. Interestingly, incubations with T09 reduced 12-HETE formation (Figure 2B) and 12-LOX protein levels (Figure 2C). However, concentrations of 12-HETE were low and originated from platelets present in cell preparations due to unavoidable platelet contaminations [39]. Again, neither elevated 15-HETE nor changed 15-LOX abundance were observed after T09 incubation in untreated macrophages. 15-LOX-2, which also catalyzes the formation of 15-HETE among other oxylipins, is found in low concentration in untreated cells, and its abundance is not stimulated by T09. While all macrophage phenotypes have comparable levels of LXRα (Figure 2C), only incubations of M2-like macrophages with the LXR agonist T09 specifically increase 15-LOX abundance and its oxylipins. No other major effect of T09 on macrophage markers, namely IL-1RII and PPARγ in M2-like and TLR2 and TLR4 in M1-like macrophages, was observed (Figure 2C). Surface markers CD14, CD86, and CD163 (Figure 2D) were not changed after T09 stimulation, suggesting T09 has no effect on specific surface receptors characteristic of macrophage polarization.

EPA- or DHA-derived mono-hydroxylated oxylipins 12-HEPE, 15-HEPE, 14-HDHA, 7-HDHA, and 17-HDHA were not affected by T09 stimulation (Figure S2, Table 1). Therefore, LXR agonist T09 specifically effects 15-LOX in M2-like macrophages.

This effect on 15-LOX in M2-like macrophages was studied more detailed in terms of the time- and dose-dependent LXR activation by T09 (Figure 3). The highest 15-LOX concentration (2.8-fold) was after a 24-h incubation (Figure 3A, Table 1). An increase in oxylipin concentrations was detected up to 24 or 30 h (Figure 3A, Table 1). Overall, the LXR-induced relative increase in oxylipin formation is more pronounced for multi-hydroxylated oxylipins (3- to 6-fold increase), such as ARA-derived 5,15-diHETE (Figure 3A), 5,12-diHETE, and 8,15-diHETE, EPA-derived 5,15-diHEPE, or DHA-derived 7,17-diHDHA (Figure S3A), than for the ARA-derived mono-hydroxylated oxylipins (2.5-fold increase) (Figure 3A, Table 1) and EPA, or DHA-derived mono-hydroxylated oxylipins (1.5-fold increase) (Figure S3A, Table 1). However, when considering the absolute amount of formed oxylipins, the levels of mono-hydroxylated oxylipins are several times higher: 184 pmol/mg protein for ARA-derived 15-HETE vs. 9.18 pmol/mg protein for 5,15-diHETE, and 45.4 pmol/mg protein for DHA-derived 17-HDHA vs. 10.7 pmol/mg protein for 7,17-diHETE (Table 1).

A strong LXR-induced dose-dependent ALOX15 abundance in M2-like macrophages was observed after 24 h of incubation (Figure 3A). At 1 μM T09, 15-LOX abundance can be maximally increased 2–3-fold; higher concentrations led to no further increase in the abundance of 15-LOX (Figure 3B). The mono-hydroxylated oxylipins show a 2–4-fold higher concentration (Figures 3B and S3B), whereas the concentration of the multi-hydroxylated metabolites is increased up to 15-fold (Figures 3B and S3B). No saturation could be achieved in the dose-dependent stimulation of M2-like cells; higher T09 concentrations were found to be cytotoxic (Figure S5). Nevertheless, incubations with 1 μM T09 reached maximum levels of 15-LOX and oxylipins, higher than all other reports about this enzyme and its products, even compared to IL-4-induced macrophages [34,40].

So far, the presence of the cytokines IL-4 and IL-13 has been required to induce ALOX15 at the mRNA and protein levels in human macrophages in cell culture experiments [34,41]. Even with prolonged incubations with 100 ng/mL LPS for more than 16 h, the 15-LOX abundance at the mRNA as well as protein level remain unchanged, whereas the 15-LOX-derived oxylipins increase by a factor of 1.5–2 [40]. In addition, the concentrations of multi-hydroxylated oxylipins, such as 5,15-diHETE and 7,17-diHDHA, were at most increased by factor 2 [40]. The LXR-mediated effect is stronger than the previously described results on 15-LOX activity in macrophages and thus could be of high biological relevance in the regulation of inflammatory pathways in macrophages. However, other pathways may also be involved in the regulation of 15-LOX abundance and activity.

15-LOX is involved in the regulation of inflammation and is thought to play a protective role in arthritis, promote wound healing and host defense and counteract fibrosis [42–44]. This is supported on the one hand by experiments of ALOX15 silencing in various experimental models, which was associated with inflammation and tissue damage [42]. On the other hand, overexpression of human reticulocyte 15-LOX in experimental models protected transgenic animals against atherosclerosis [45].

Anti-inflammatory properties have been described for several of the 15-LOX products. ARA-derived 15-HETE can activate PPARγ and inhibit neutrophil migration, degranulation, and superoxide formation [17,45–49]. 15-HETE can also reduce inflammatory signaling by regulating the TNFα mRNA half-life [50,51]. EPA-derived 15-HEPE is also thought to have anti-inflammatory properties. In transgenic fat1 mice expressing an n3-PUFA generating desaturase, a protective effect was observed after DSS-induced colitis, which was attributed to the high concentrations of 12-HEPE, the major 15-LOX metabolite of EPA in mice. The results were corroborated by silencing of ALOX15 in fat1 transgenic mice, where 12-HEPE concentration was strongly reduced. Moreover, the 15-LOX metabolite 15-HEPE was also reduced in ALOX15-silenced fat1 transgenic mice, and the wild-type mice were protected against the development of DSS-induced colitis after intraperitoneal injections with 15-HEPE [35].

Upon interaction with several LOX enzymes, 5,15-diHETE or 8,15-diHETE can be formed, which are associated with neutrophil and eosinophil chemotactic activity [46,52,53]. EPA-derived 5,15-diHEPE, also known as resolvin E4, has been described to induce efferocytosis of apoptotic neutrophils in human macrophages and senescent red blood cells along with an upregulation of PPARγ gene expression, leading to a resolution of inflammation [54]. In addition, the ALOX15 DHA-derived multi-hydroxylated metabolite 7,17-diHDHA, also called resolvin D5, increases phagocytic activity in neutrophils and macrophages and decreases the formation of pro-inflammatory mediators such as TNFα and NF-κB [55,56]. However, these multi-hydroxylated oxylipins are synthesized in cells and tissues only at low concentrations, and many of the proposed effects and signaling pathways are a matter of discussion [30].

Here, we show that LXR agonist T09 specifically elevates ALOX15 abundance, while the effects on markers and proteins involved in the regulation of inflammation (PPARγ, TLR2, and TLR4 or COX) were minor. This suggests a trend towards a more anti-inflammatory biology in the macrophages due to the increased abundance and activity of 15-LOX. Thus, with the correlation between LXR activation and the formation of oxylipins, we conclude that part of the anti-inflammatory effects of LXR [16,20,31] are mediated by 15-LOX and its products. Given that the effects of 15-LOX modulation are more pronounced compared to the TLR4 ligand LPS, this argues that LXR seems to be a major regulatory pathway in macrophages. However, other pathways might also contribute to this effect, as T09 also has a side activity on other nuclear receptors.

We examined the specificity of the ALOX15 regulation by LXR activation. We analyzed the effects of the related nuclear receptors FXR and RXR (Figure 4) with regard to 15-LOX modulation in M2-like macrophages using specific agonists and antagonists. Incubations with the LXR antagonist GSK2033 did not lead to changes in 15-LOX abundance or the formation of its metabolites (Figure 5), indicating that LXR might not be the only factor involved in ALOX-15 regulation.

Following treatment of M2-like cells with the FXR agonist hyodeoxycholic acid (HDCA) [57], the 15-LOX abundance remains unchanged. The FXR agonist fexaramine leads to a 2-fold increase in 15-LOX abundance (Figure 5 Proteomics), and the corresponding oxylipin concentrations are also increased by 2 to 5-fold (Figure 5 Oxylipins). However, due to the rather nonspecific binding pocket [58,59] of the nuclear receptors, the binding of ligands overlaps. Thus, the effect of fexaramine could be caused by a side activity on LXR, though it was previously described that fexaramine cannot activate LXR [60]. Consistently, it has been shown that the RXR agonists bexarotene and 9-*cis* retinoic acid (9-RA) have agonistic activity toward LXR [61], and thus, bexarotene increased 15-LOX abundance 2-fold (Figure 5 Proteomics) and its products 2- to 5-fold (Figure 5 Oxylipins). These results also might indicate the involvement of other nuclear receptors or an off-target effect of the substances on LXR. 9-RA showed lower effects on 15-LOX: the 15-LOX concentrations apparently remained unchanged while 12-HETE and 5,15-diHETE formation increased (Figure 5).

Finally, we searched for endogenous ligands eliciting the LXR-mediated effects on the ARA cascade in M2-like macrophages (Figure 4D). Oxysterols and metabolites from cholesterol biosynthesis have been previously described as LXR activators [22,62–64].

Treatment with the autoxidatively formed 7-keto cholesterol leads to a moderate increase in 15-LOX abundance and activity (Figure 5). The 24(*S*),25-Ep cholesterol elevates 15-LOX products in a concentration-independent manner (2-fold) (Figure 5 Oxylipins), whereas only incubations with 10 μM 24(*S*),25-Ep cholesterol increase mono-hydroxylated oxylipins 3-fold and multi-hydroxylated oxylipins 5-fold (Figure 5 Oxylipins). For 24(*S*)-OH cholesterol and 25-OH cholesterol, a weaker effect was observed as they moderately increased LOX products, while no difference in 15-LOX abundance was observed (Figure 5 Proteomics). For 24(*S*)-OH cholesterol, a trend towards increased 15-LOX-derived oxylipin concentrations (up to 3-fold) is observed (Figure 5 Oxylipins). In incubations with 25-OH cholesterol, the oxylipin concentrations were reduced by half (Figure 5 Oxylipins). 22(*R*)-OH cholesterol has the strongest effect on 15-LOX activity of all the oxysterols studied. While the mono-hydroxylated oxylipins are increased 2-fold, the increase of the multi-hydroxylated oxylipins is up to 15-fold (Figure 5 Oxylipins), together with a 2-fold increase in 15-LOX abundance.

The activity of oxysterols on ALOX15 in M2-like macrophages is consistent with ligand binding affinity to the LXR [21,63]. In structure/activity studies, it was shown that the position of the functional group is of great importance, the essential positions being C-22 and C-24 [63,65]. In our experiments, we were also able to correlate the effect strength with the position of the functional group. Thus, when incubated with 22(*R*)-OH cholesterol and 24(*S*),25-Ep cholesterol, we observed the strongest effect on 15-LOX abundance and activity, whereas with 25-OH cholesterol, hardly any changes were observed. Our results are consistent with the effects described, where a significant LXR activation is observed for 22(*R*)-OH cholesterol > 24(*S*)-OH cholesterol > 25-OH cholesterol, whereas 25-OH cholesterol has little to no effect [21,63,64,66]. Other significantly important ligands are described to be 24(*S*),25-Ep cholesterol, 20(*S*)-OH cholesterol, and 27-OH cholesterol [63,64,67]. Of note, 24(*S*),25-Ep cholesterol is not formed from cholesterol like other oxysterols but is a by-product of cholesterol biosynthesis (mevalonate pathway) [63,67,68]. In addition, the central intermediate from cholesterol biosynthesis, desmosterol, caused strong effects on 15-LOX. Following stimulation with desmosterol, 15-LOX is overexpressed almost 3-fold, which is comparable to the effect of T09 on 15-LOX abundance (Figure 5 Proteomics). 15-LOX activity in incubations with desmosterol and T09 is also comparable, with the mono-hydroxylated oxylipins also being increased 4-fold and the multi-hydroxylated oxylipins up to 16-fold (Figure 5 Oxylipins). Thus, we have found active endogenous LXR ligands, namely desmosterol, 22(*R*)-OH cholesterol and 24(*S*),25-Ep cholesterol, that are able to massively increase 15-LOX abundance and activity through LXR.

Oxysterols occur in mammalian tissues in micromolar concentrations and circulate in only nanomolar concentrations [63,67,69]. Elevated (circulating) oxysterol levels are associated with pathological structures, such as foam cells or atherosclerotic lesions [32,67,70,71]. High 24(*S*)-OH cholesterol plasma levels are associated with Alzheimer's disease [67,72,73]. Desmosterol can be found accumulated mainly in macrophage foam cells and atherosclerotic plaques, where it regulates via LXR the activation of genes involved in cholesterol efflux [74]. Due to the induction of ALOX15 abundance and the elevated formation of anti-inflammatory oxylipins, desmosterol might also be involved in the reduction of atherosclerotic lesions via LXR activation. 22(*R*)-OH cholesterol, 24(*S*)-OH cholesterol, and 24(*S*),25-Ep cholesterol have been shown to inhibit inflammatory signaling in cell culture models [75]. As many oxysterols and metabolites of cholesterol biosynthesis are involved in the regulation of inflammatory and immune responses [16,76–78], this study of the effect of cholesterol derivatives on macrophages is of great importance to better understand the interactions between sterol metabolism and LXR-mediated lipid mediators. The induction of ALOX15 in macrophages is associated with anti-inflammatory properties [45]. Our results show that LXR upregulation may be related to the modulation of 15-LOX and the formation of anti-inflammatory oxylipins. Thus, the observed relationship between LXR and oxylipin formation activated by endogenous sterols may represent an important feedback regulation in macrophages, where sterols indirectly modulate 15-LOX through their binding to LXR, initiating inflammatory resolution.

Limitations: As with all mechanistic cell culture studies using cell model systems and chemical probes, our study has limitations: The primary macrophages used, derived from different subjects, showed different (basal) levels of 15-LOX abundance and activity, which could be a confounder in the observed regulation by LXR agonists. Macrophages derived from healthy donors from a blood donation center, but no detailed information about their health status, sex, age, or medications were available. These factors might also contribute to the regulation of 15-LOX. Regarding the chemical probes, it should be noted that one agonist and one antagonist of LXR and no antagonists of FXR and RXR could be tested in the study, and they may elicit non-specific effects at the tested concentrations.

4. Materials and Methods

4.1. Chemicals

Human AB plasma was obtained from the blood donor service of the University Hospital of Düsseldorf (Düsseldorf, Germany). Lymphocyte separation medium 1077 was from PromoCell (Heidelberg, Germany). Recombinant human colony stimulating factors M-CSF and GM-CSF, IFNγ, and IL-4 produced in *Escherichia coli* were purchased from PeproTech Germany (Hamburg, Germany). RPMI 1640 cell culture medium, L-glutamine and penicillin/streptomycin (5.000 units penicillin and 5 mg streptomycin/mL), lipopolysaccharide (LPS) from *E. coli* (0111:B4, product number: L2630), dextran from *Leuconostoc* spp. (molecular weight 450,000–650,000), copper sulfate pentahydrate, iodoacetamide, dimethylsulfoxide (DMSO), and desmosterol were obtained from Sigma (Schnellendorf, Germany). Trypsin (>6000 U/g, from porcine pancreas), protease-inhibitor mix M (AEBSF, Aprotinin, Bestatin, E-64, Leupeptin, Pepstatin A), and resazurin were purchased from SERVA Electrophoresis GmbH (Heidelberg, Germany). Ammonium hydrogen carbonate, sodium deoxycholate, urea, and formaldehyde were from Carl Roth (Karlsruhe, Germany).

Acetonitrile, methanol (LC-MS grade), acetone (HPLC grade), acetic acid (Optima LC-MS grade), and BCA reagent A were purchased from Fisher Scientific (Schwerte, Germany). Ethyl acetate (HPLC grade) and n-hexane (HPLC grade) were bought from VWR (Darmstadt, Germany). The ultra-pure water with a conductivity of >18 MΩ·cm was generated by the Barnstead Genpure Pro system from Thermo Fisher Scientific (Langenselbold, Germany). Oxylipin standards and deuterated oxylipin standards used as internal standards and tested compounds (T0901317 (T09), 22(*R*)-OH cholesterol, 24(*S*)-OH cholesterol, 25-OH cholesterol, 24(*S*),25-Ep cholesterol, 7-keto cholesterol, GSK2033, fexaramine, hyodeoxycholic acid, bexarotene, and 9-*cis* retinoic acid) were purchased from Cayman Chemical (local distributor Biomol, Hamburg, Germany). Unlabeled and heavy labeled (lys: uniformly labeled (U)-$^{13}C_6$; U-$^{15}N_2$; arg: U-$^{13}C_6$; U-$^{15}N_4$) peptide standards were purchased from JPT Peptides (Berlin, Germany).

4.2. Cultivation of Macrophages from Human PBMC

Primary human macrophages were purified and differentiated as described [40]. Buffy coats were obtained from the generation of erythrocyte concentrates from the blood donor services at the University Hospital of Düsseldorf, Germany, with the informed consent of healthy human subjects. The study was approved by the Ethical Committee of the University of Wuppertal. In brief, primary human blood monocytic cells (PBMC) were isolated from fresh buffy coats by dextran (5%) sedimentation for 45 min. The supernatant was layered on lymphocyte separation medium and centrifuged for 10 min at 1000× *g* without deceleration. PBMC sedimented on top of the lymphocyte separation medium was collected. Following washing with PBS twice, cell pellets were resuspended in RPMI 1640 medium supplemented with 1% L-glutamine and 1% penicillin/streptomycin (P/S). Cell suspensions were transferred to petri dishes and incubated for 1 h (37 °C; 5% CO_2; humidified atmosphere). Afterwards, dishes were washed twice with RPMI 1640 medium supplemented with 1% L-glutamine and 1% P/S to remove non-adherent cells and layered with RPMI 1640 growth medium, which contains 2 g/L glucose and is supplemented with 1% L-glutamine, 1% P/S, and 5% heat-inactivated human AB plasma. The monocytes were differentiated into different macrophage phenotypes for 8 days. For the M1-like phenotype, the growth medium was supplemented with 10 ng/mL GM-CSF, and for the M2-like phenotype, with 10 ng/mL M-CSF. The growth medium was refreshed every other day. In addition, for the last 48 h the cells were incubated with 10 ng/mL IFNγ (M1) or 10 ng/mL IL-4 (M2). Six hours before the harvest, the M1-like cells were additionally stimulated with 100 ng/mL LPS. For the untreated phenotype, cells were cultivated in growth medium without the addition of cytokines.

The differentiated macrophages were treated with test compounds at different time points and concentrations.

The cytotoxic effect of the test substances was investigated by means of the resazurin (alamar blue) assay [79], and only non-toxic concentrations were used (Figure S5).

The cells were harvested using the cold shock method by washing with PBS and incubating in ice cold PBS/EDTA (20 min/4 °C). The cells were collected by scraping, centrifugation, and pelleting followed by washing with PBS containing protease inhibitors. The harvested primary macrophage pellets were frozen at −80 °C until use.

4.3. Analysis of Oxylipins and Protein Levels by LC-MS/MS

The analysis of oxylipins and protein levels was carried out from the same cell pellet. Cell pellets were resuspended in PBS containing an antioxidant/inhibitor mixture (0.2 mg/mL BHT, 100 µM indomethacin, 100 µM *trans*-4-(-4-(3-adamantane-1-yl-ureido-9-cyclohexyloxy)-benzoic acid (t-AUCB) in MeOH) and sonicated [80,81]. The protein content was determined by the bicinchoninic acid (BCA) assay. Following addition of 10 µL of the oxylipin internal standards ($^2H_4$-PGE$_2$, $^2H_4$-PGD$_2$, $^2H_4$-TxB$_2$, $^2H_4$-13,14-dihydro-15-keto PGE$_2$, $^2H_4$-15-deoxy-$^{\Delta 12,14}$-PGJ$_2$, $^2H_4$-6-keto-PGF$_{1\alpha}$, $^2H_4$-8-*iso*-PGF$_{2\alpha}$, $^2H_4$-PGE$_2$, $^2H_4$-PGD$_2$, $^2H_4$-TxB$_2$, $^2H_4$-PGF$_{2\alpha}$, $^2H_5$-RvD2, $^2H_{11}$-8,12-*iso*-iPF$_{2\alpha}$-VI, $^2H_5$-LxA$_4$, $^2H_5$-RvD1, $^2H_4$-PGB$_2$, $^2H_4$-LTB$_4$, $^2H_4$-9,10-DiHOME, $^2H_{11}$-11,12-DiHETrE, $^2H_6$-20-HETE, $^2H_4$-13-HODE, $^2H_4$-9-HODE, $^2H_8$-15-HETE, $^2H_3$-13-oxoODE, $^2H_8$-12-HETE, $^2H_8$-5-HETE, $^2H_4$-12(13)-EpOME, $^2H_{11}$-14(15)-EpETrE, $^2H_7$-5-oxoETE, $^2H_{11}$-8(9)-EpETrE, each 100 nM in MeOH) and protein precipitation in methanol (−80 °C; 30 min), the supernatant was used for the oxylipin analysis whereas the protein pellet was frozen at −80 °C for a later protein level analysis [39].

The oxylipin analysis was carried out as described [80,81]. In brief, the oxylipins were extracted from the supernatants using Bond Elut Certify II SPE cartridges (200 mg, 3 mL, Agilent, Waldbronn, Germany). Following elution of the oxylipins using ethyl acetate/n hexane/acetic acid (75/25/1, *v*/*v*/*v*), samples were evaporated (vacuum concentrator, 30 °C, 1 mbar; Christ, Osterode, Germany), and the residue was reconstituted in 50 µL internal standard 2 (1-(1-(ethylsulfonyl)piperidin-4-yl)-3-(4-(trifluoromethoxy)-phenyl)-urea, 12-(3-adamantan-1-yl-ureido)-dodecanoic acid, 12-oxo-phytodienoic acid, and aleuritic acid) [80,81]. Oxylipins were analyzed using an 1290 Infinity II System (Agilent) coupled to a 5500 QTRAP instrument (Sciex, Darmstadt, Germany) in ESI(−)-mode operated in scheduled selected reaction monitoring [82].

The analysis of the protein levels was carried out as described [39,83]. In brief, the protein pellet was dissolved in 5% (*w*/*v*) sodium deoxycholate with protease inhibitor (100/1; *v*/*v*), precipitated with ice-cold acetone (4 volumes), and centrifuged (4 °C, 15,000× *g*, 20 min). Following subsequent incubations with 200 mM dithiothreitol (in 50 mM NH_4HCO_3), 200 mM iodoacetamide (in 50 mM NH_4HCO_3), and again 200 mM dithiothreitol, the proteins were digested overnight (15 h) using 100 µg/mL trypsin in 50 mM acetic acid (trypsin-to-protein ratio of 1:50). The reaction was stopped by acidification with concentrated acetic acid (pH 3–4). Following the addition of internal standards (heavy labeled peptides with sequences corresponding to each analyte peptide), the samples were extracted using Strata-X SPE 33 µm Polymeric Reversed Phase cartridges (100 mg/3 mL, Phenomenex LTD, Aschaffenburg, Germany). The peptides were eluted with 70% acetonitrile/0.1% acetic acid, evaporated, and finally reconstituted in 15% acetonitrile/0.1% acetic acid. The peptides were analyzed on an 1290 Infinity II System (Agilent) coupled to a 6500+ QTRAP instrument (Sciex) in ESI(+)-mode operated in scheduled selected reaction monitoring mode [83].

MS data analysis was performed with the software MultiQuant^{TM} 3.0.2 (Sciex) using a Gaussian smooth width of 1 and the MQ4 integration algorithm. Concentrations were calculated via external calibration using internal standards based on the ratio of the analyte and internal standard areas. The quantified peptide/protein and oxylipin concentrations were normalized to the absolute protein content determined via the BCA assay.

4.4. Immunofluorescence Labelling

PBMC were seeded in 24-well plates on coverslips (Sarstedt, Nümbrecht, Germany) and cultivated as described above. For labeling, polarized macrophages were fixed in 4% formaldehyde in PBS for 15 min at 37 °C. Following washing with TBS the cells were permeabilized and blocked with TBS, containing 0.3% Tween 20 and 1% bovine serum albumin (BSA) for 30 min at 30 °C. The primary macrophages were then incubated overnight with the primary antibodies 1:100 diluted in TBS/0.3% Tween 20/1% BSA against CD14, CD86, and CD163 (all mouse anti-human; Bio-Rad Laboratories, Feldkirchen, Germany). The coverslips were washed extensively with TBS/0.3% Tween 20 and then incubated with FITC-conjugated secondary antibody (1:300 in TBS/0.3% Tween 20/1% BSA; goat anti-mouse IgG Alexa Fluor 488, Thermo Scientific) for 45 min at 30 °C and again extensively washed with TBS/0.3% Tween 20. The nuclei were stained with 10 µg/mL Hoechst 33258 (Sigma) for 45 s and finally the coverslips were observed using a fluorescence microscope (Leica DM6 B, Wetzlar, Germany) [84].

4.5. Statistical Analysis

Statistical analysis was performed using GraphPad Prism (GraphPad Software Inc. version 6.01, San Diego, CA, USA). Data are presented as mean ± standard error of mean (SEM). Statistical analyses were performed by a two-tailed, unpaired student's *t*-test. Differences were considered significant at *p*-values < 0.05 or 0.01.

5. Conclusions

We demonstrate that LXR activation by T09 exerts dramatic effects on the ARA cascade, specifically in M2-like macrophages, by increasing the abundance of 15-LOX and increasing the formation of its products. No or only a moderate effect of T09 was observed on untreated and M1-like macrophages, as well as surface markers and PPARγ, IL-1RII, TLR2, and TLR4 abundance.

The investigated agonists of FXR and RXR only had a moderate effect on 15-LOX abundance and activity compared to T09, presumably mediated by LXR through side activity.

We found that sterols are important endogenous LXR agonists regulating ALOX15. We could show that particularly the oxysterols 24(*S*),25-Ep cholesterol, and 22(*R*)-OH cholesterol and the cholesterol precursor desmosterol modulate the ARA cascade in macrophages, presumably by LXR. Thus, our results show a new cross-link between two lipid mediator classes: sterols/oxysterols and oxylipins. Sterols affect the LXR-mediated formation of 15-LOX-derived oxylipins. This regulatory mechanism should be further investigated, as the 15-LOX-derived oxylipins are suggested to play an important role in inflammatory resolution.

A detailed characterization of this regulatory pathway could provide insights into the lipid mediator switch in macrophages leading to the formation of anti-inflammatory oxylipins.

Supplementary Materials: The following supporting information can be downloaded at: https://www.mdpi.com/article/10.3390/molecules29081745/s1, Table S1: Selection of iNOS peptides from an in silico tryptic digest; Table S2: Selected proteotypic peptides (PTPs) for LXRα, LXRβ, PPARγ, IL-1RII, TLR2, TLR4, and iNOS; Table S3: Selected transitions for (A) unlabeled and (B) heavy labeled (lys: U-$^{13}C_6$; U-$^{15}N_2$; arg: U-$^{13}C_6$; U-$^{15}N_4$) peptide data for LXRα, LXRβ, PPARγ, IL-1RII, TLR2, TLR4, iNOS; Table S4: Oxylipin concentrations and 15-LOX levels in M2-like macrophages induced by LXR, FXR, and RXR agonists and cholesterol precursors and metabolites; Figure S1: Immunostaining of macrophage surface markers; Figure S2: Effect of the LXR agonist T09 on the monohydroxylated oxylipin formation; Figure S3: Time- and dose-dependent 15-LOX activity; Figure S4: 15-LOX activity induced by LXR, FXR, and RXR agonists and identification of cholesterol precursors and metabolites; Figure S5: Cell viability assay of the test compounds in M2-like macrophages. Refs. [85–96] are cited in Supplementary Materials.

Author Contributions: Conceptualization, M.M. and N.H.S.; methodology, M.M. and N.M.H.; validation, M.M.; formal analysis, M.M.; investigation, M.M., R.O., R.K. and N.M.H.; data curation, M.M. and R.O.; writing—original draft preparation, M.M. and N.H.S.; writing—review and editing, R.O., R.K. and N.H.S.; visualization, M.M. and R.O.; supervision, N.H.S.; project administration, N.H.S.; funding acquisition, N.H.S. All authors have read and agreed to the published version of the manuscript.

Funding: This research was funded by the German Research Foundation (DFG), grant number SCHE 1801.

Institutional Review Board Statement: The study was conducted in accordance with the Declaration of Helsinki and approved by the Ethics Committee of the University of Wuppertal for studies involving humans.

Informed Consent Statement: Informed consent was obtained from all subjects involved in the study.

Data Availability Statement: Data are contained within the article and Supplementary Materials.

Conflicts of Interest: The authors declare no conflicts of interest.

References

1. Qu, X.; Tang, Y.; Hua, S. Immunological Approaches Towards Cancer and Inflammation: A Cross Talk. *Front. Immunol.* **2018**, *9*, 563. [CrossRef] [PubMed]
2. Nathan, C. Points of control in inflammation. *Nature* **2002**, *420*, 846–852. [CrossRef] [PubMed]
3. Medzhitov, R. Origin and physiological roles of inflammation. *Nature* **2008**, *454*, 428–435. [CrossRef] [PubMed]
4. Kasahara, T.; Matsushima, K. Macrophage signaling, apoptosis, lectins and leukocyte trafficking. *Trends Immunol.* **2001**, *22*, 593–594. [CrossRef] [PubMed]
5. Fujiwara, N.; Kobayashi, K. Macrophages in inflammation. *Curr. Drug Targets Inflamm. Allergy* **2005**, *4*, 281–286. [CrossRef] [PubMed]
6. Ortega-Gomez, A.; Perretti, M.; Soehnlein, O. Resolution of inflammation: An integrated view. *EMBO Mol. Med.* **2013**, *5*, 661–674. [CrossRef]
7. Gately, M.K.; Renzetti, L.M.; Magram, J.; Stern, A.S.; Adorini, L.; Gubler, U.; Presky, D.H. The interleukin-12/interleukin-12-receptor system: Role in normal and pathologic immune responses. *Annu. Rev. Immunol.* **1998**, *16*, 495–521. [CrossRef] [PubMed]
8. Rot, A.; von Andrian, U.H. Chemokines in innate and adaptive host defense: Basic chemokinese grammar for immune cells. *Annu. Rev. Immunol.* **2004**, *22*, 891–928. [CrossRef] [PubMed]
9. Tarique, A.A.; Logan, J.; Thomas, E.; Holt, P.G.; Sly, P.D.; Fantino, E. Phenotypic, functional, and plasticity features of classical and alternatively activated human macrophages. *Am. J. Respir. Cell Mol. Biol.* **2015**, *53*, 676–688. [CrossRef]
10. Fleetwood, A.J.; Lawrence, T.; Hamilton, J.A.; Cook, A.D. Granulocyte-macrophage colony-stimulating factor (CSF) and macrophage CSF-dependent macrophage phenotypes display differences in cytokine profiles and transcription factor activities: Implications for CSF blockade in inflammation. *J. Immunol.* **2007**, *178*, 5245–5252. [CrossRef]
11. Nathan, C.F.; Murray, H.W.; Wiebe, M.E.; Rubin, B.Y. Identification of interferon-gamma as the lymphokine that activates human macrophage oxidative metabolism and antimicrobial activity. *J. Exp. Med.* **1983**, *158*, 670–689. [CrossRef] [PubMed]
12. Stein, M.; Keshav, S.; Harris, N.; Gordon, S. Interleukin 4 potently enhances murine macrophage mannose receptor activity: A marker of alternative immunologic macrophage activation. *J. Exp. Med.* **1992**, *176*, 287–292. [CrossRef] [PubMed]

13. Koh, T.J.; DiPietro, L.A. Inflammation and wound healing: The role of the macrophage. *Expert. Rev. Mol. Med.* **2011**, *13*, e23. [CrossRef]
14. Fadok, V.A.; Bratton, D.L.; Konowal, A.; Freed, P.W.; Westcott, J.Y.; Henson, P.M. Macrophages that have ingested apoptotic cells in vitro inhibit proinflammatory cytokine production through autocrine/paracrine mechanisms involving TGF-beta, PGE2, and PAF. *J. Clin. Investig.* **1998**, *101*, 890–898. [CrossRef]
15. Michlewska, S.; Dransfield, I.; Megson, I.L.; Rossi, A.G. Macrophage phagocytosis of apoptotic neutrophils is critically regulated by the opposing actions of pro-inflammatory and anti-inflammatory agents: Key role for TNF-alpha. *FASEB J.* **2009**, *23*, 844–854. [CrossRef] [PubMed]
16. Roszer, T. Transcriptional control of apoptotic cell clearance by macrophage nuclear receptors. *Apoptosis* **2017**, *22*, 284–294. [CrossRef] [PubMed]
17. Huang, J.T.; Welch, J.S.; Ricote, M.; Binder, C.J.; Willson, T.M.; Kelly, C.; Witztum, J.L.; Funk, C.D.; Conrad, D.; Glass, C.K. Interleukin-4-dependent production of PPAR-gamma ligands in macrophages by 12/15-lipoxygenase. *Nature* **1999**, *400*, 378–382. [CrossRef]
18. Roszer, T.; Menendez-Gutierrez, M.P.; Lefterova, M.I.; Alameda, D.; Nunez, V.; Lazar, M.A.; Fischer, T.; Ricote, M. Autoimmune kidney disease and impaired engulfment of apoptotic cells in mice with macrophage peroxisome proliferator-activated receptor gamma or retinoid X receptor alpha deficiency. *J. Immunol.* **2011**, *186*, 621–631. [CrossRef] [PubMed]
19. Schulman, I.G. Liver X receptors link lipid metabolism and inflammation. *FEBS Lett.* **2017**, *591*, 2978–2991. [CrossRef]
20. Noelia, A.; Bensinger, S.J.; Hong, C.; Beceiro, S.; Bradley, M.N.; Zelcer, N.; Deniz, J.; Ramirez, C.; Diaz, M.; Gallardo, G.; et al. Apoptotic cells promote their own clearance and immune tolerance through activation of the nuclear receptor LXR. *Immunity* **2009**, *31*, 245–258.
21. Edwards, P.A.; Kennedy, M.A.; Mak, P.A. LXRs; oxysterol-activated nuclear receptors that regulate genes controlling lipid homeostasis. *Vascul. Pharmacol.* **2002**, *38*, 249–256. [CrossRef] [PubMed]
22. Hong, C.; Tontonoz, P. Coordination of inflammation and metabolism by PPAR and LXR nuclear receptors. *Curr. Opin. Genet. Dev.* **2008**, *18*, 461–467. [CrossRef] [PubMed]
23. Repa, J.J.; Mangelsdorf, D.J. The role of orphan nuclear receptors in the regulation of cholesterol homeostasis. *Annu. Rev. Cell Dev. Biol.* **2000**, *16*, 459–481. [CrossRef] [PubMed]
24. Kim, K.H.; Lee, G.Y.; Kim, J.I.; Ham, M.; Won Lee, J.; Kim, J.B. Inhibitory effect of LXR activation on cell proliferation and cell cycle progression through lipogenic activity. *J. Lipid Res.* **2010**, *51*, 3425–3433. [CrossRef] [PubMed]
25. Steffensen, K.R.; Gustafsson, J.A. Putative metabolic effects of the liver X receptor (LXR). *Diabetes* **2004**, *53* (Suppl. S1), S36–S42. [CrossRef] [PubMed]
26. Martinez, F.O.; Gordon, S. The M1 and M2 paradigm of macrophage activation: Time for reassessment. *F1000Prime Rep.* **2014**, *6*, 13. [CrossRef] [PubMed]
27. Serhan, C.N.; Chiang, N.; Dalli, J.; Levy, B.D. Lipid mediators in the resolution of inflammation. *Cold Spring Harb. Perspect. Biol.* **2014**, *7*, a016311. [CrossRef] [PubMed]
28. Buckley, C.D.; Gilroy, D.W.; Serhan, C.N. Proresolving lipid mediators and mechanisms in the resolution of acute inflammation. *Immunity* **2014**, *40*, 315–327. [CrossRef] [PubMed]
29. Kahnt, A.S.; Schebb, N.H.; Steinhilber, D. Formation of lipoxins and resolvins in human leukocytes. *Prostaglandins Other Lipid Mediat.* **2023**, *166*, 106726. [CrossRef]
30. Schebb, N.H.; Kuhn, H.; Kahnt, A.S.; Rund, K.M.; O'Donnell, V.B.; Flamand, N.; Peters-Golden, M.; Jakobsson, P.J.; Weylandt, K.H.; Rohwer, N.; et al. Formation, Signaling and Occurrence of Specialized Pro-Resolving Lipid Mediators-What is the Evidence so far? *Front. Pharmacol.* **2022**, *13*, 838782. [CrossRef]
31. Snodgrass, R.G.; Benatzy, Y.; Schmid, T.; Namgaladze, D.; Mainka, M.; Schebb, N.H.; Lutjohann, D.; Brune, B. Efferocytosis potentiates the expression of arachidonate 15-lipoxygenase (ALOX15) in alternatively activated human macrophages through LXR activation. *Cell Death Differ.* **2021**, *28*, 1301–1316. [CrossRef] [PubMed]
32. Shibata, N.; Glass, C.K. Macrophages, oxysterols and atherosclerosis. *Circ. J.* **2010**, *74*, 2045–2051. [CrossRef] [PubMed]
33. Ivanov, I.; Kuhn, H.; Heydeck, D. Structural and functional biology of arachidonic acid 15-lipoxygenase-1 (ALOX15). *Gene* **2015**, *573*, 1–32. [CrossRef] [PubMed]
34. Kuhn, H.; Banthiya, S.; van Leyen, K. Mammalian lipoxygenases and their biological relevance. *Biochim. Biophys. Acta* **2015**, *1851*, 308–330. [CrossRef]
35. Rohwer, N.; Chiu, C.Y.; Huang, D.; Smyl, C.; Rothe, M.; Rund, K.M.; Helge Schebb, N.; Kuhn, H.; Weylandt, K.H. Omega-3 fatty acids protect from colitis via an Alox15-derived eicosanoid. *FASEB J.* **2021**, *35*, e21491. [CrossRef]
36. Gordon, S.; Martinez, F.O. Alternative activation of macrophages: Mechanism and functions. *Immunity* **2010**, *32*, 593–604. [CrossRef] [PubMed]
37. Mantovani, A.; Sica, A.; Sozzani, S.; Allavena, P.; Vecchi, A.; Locati, M. The chemokine system in diverse forms of macrophage activation and polarization. *Trends Immunol.* **2004**, *25*, 677–686. [CrossRef]
38. Xue, J.; Schmidt, S.V.; Sander, J.; Draffehn, A.; Krebs, W.; Quester, I.; De Nardo, D.; Gohel, T.D.; Emde, M.; Schmidleithner, L.; et al. Transcriptome-based network analysis reveals a spectrum model of human macrophage activation. *Immunity* **2014**, *40*, 274–288. [CrossRef]

39. Hartung, N.M.; Mainka, M.; Pfaff, R.; Kuhn, M.; Biernacki, S.; Zinnert, L.; Schebb, N.H. Development of a quantitative proteomics approach for cyclooxygenases and lipoxygenases in parallel to quantitative oxylipin analysis allowing the comprehensive investigation of the arachidonic acid cascade. *Anal. Bioanal. Chem.* **2023**, *415*, 913–933. [CrossRef]
40. Ebert, R.; Cumbana, R.; Lehmann, C.; Kutzner, L.; Toewe, A.; Ferreiros, N.; Parnham, M.J.; Schebb, N.H.; Steinhilber, D.; Kahnt, A.S. Long-term stimulation of toll-like receptor-2 and -4 upregulates 5-LO and 15-LO-2 expression thereby inducing a lipid mediator shift in human monocyte-derived macrophages. *Biochim. Biophys. Acta Mol. Cell Biol. Lipids* **2020**, *1865*, 158702. [CrossRef]
41. Snodgrass, R.G.; Brune, B. Regulation and Functions of 15-Lipoxygenases in Human Macrophages. *Front. Pharmacol.* **2019**, *10*, 719. [CrossRef] [PubMed]
42. Kronke, G.; Katzenbeisser, J.; Uderhardt, S.; Zaiss, M.M.; Scholtysek, C.; Schabbauer, G.; Zarbock, A.; Koenders, M.I.; Axmann, R.; Zwerina, J.; et al. 12/15-lipoxygenase counteracts inflammation and tissue damage in arthritis. *J. Immunol.* **2009**, *183*, 3383–3389. [CrossRef] [PubMed]
43. Gronert, K.; Maheshwari, N.; Khan, N.; Hassan, I.R.; Dunn, M.; Laniado Schwartzman, M. A role for the mouse 12/15-lipoxygenase pathway in promoting epithelial wound healing and host defense. *J. Biol. Chem.* **2005**, *280*, 15267–15278. [CrossRef] [PubMed]
44. Ogawa, M.; Ishihara, T.; Isobe, Y.; Kato, T.; Kuba, K.; Imai, Y.; Uchino, Y.; Tsubota, K.; Arita, M. Eosinophils promote corneal wound healing via the 12/15-lipoxygenase pathway. *FASEB J.* **2020**, *34*, 12492–12501. [CrossRef] [PubMed]
45. Kuhn, H.; Walther, M.; Kuban, R.J. Mammalian arachidonate 15-lipoxygenases structure, function, and biological implications. *Prostaglandins Other Lipid Mediat.* **2002**, *68–69*, 263–290.
46. Gabbs, M.; Leng, S.; Devassy, J.G.; Monirujjaman, M.; Aukema, H.M. Advances in Our Understanding of Oxylipins Derived from Dietary PUFAs. *Adv. Nutr.* **2015**, *6*, 513–540. [CrossRef] [PubMed]
47. Naruhn, S.; Meissner, W.; Adhikary, T.; Kaddatz, K.; Klein, T.; Watzer, B.; Muller-Brusselbach, S.; Muller, R. 15-hydroxyeicosatetraenoic acid is a preferential peroxisome proliferator-activated receptor beta/delta agonist. *Mol. Pharmacol.* **2010**, *77*, 171–184. [CrossRef] [PubMed]
48. Takata, S.; Papayianni, A.; Matsubara, M.; Jimenez, W.; Pronovost, P.H.; Brady, H.R. 15-Hydroxyeicosatetraenoic acid inhibits neutrophil migration across cytokine-activated endothelium. *Am. J. Pathol.* **1994**, *145*, 541–549. [PubMed]
49. Smith, R.J.; Justen, J.M.; Nidy, E.G.; Sam, L.M.; Bleasdale, J.E. Transmembrane signaling in human polymorphonuclear neutrophils: 15(S)-hydroxy-(5Z,8Z,11Z,13E)-eicosatetraenoic acid modulates receptor agonist-triggered cell activation. *Proc. Natl. Acad. Sci. USA* **1993**, *90*, 7270–7274. [CrossRef] [PubMed]
50. Serhan, C.N. Lipoxins and aspirin-triggered 15-epi-lipoxins are the first lipid mediators of endogenous anti-inflammation and resolution. *Prostaglandins Leukot. Essent. Fatty Acids* **2005**, *73*, 141–162. [CrossRef]
51. Ferrante, J.V.; Ferrante, A. Novel role of lipoxygenases in the inflammatory response: Promotion of TNF mRNA decay by 15-hydroperoxyeicosatetraenoic acid in a monocytic cell line. *J. Immunol.* **2005**, *174*, 3169–3172. [CrossRef]
52. Morita, E.; Schroder, J.M.; Christophers, E. Identification of a novel and highly potent eosinophil chemotactic lipid in human eosinophils treated with arachidonic acid. *J. Immunol.* **1990**, *144*, 1893–1900. [CrossRef] [PubMed]
53. Powell, W.S.; Gravel, S.; MacLeod, R.J.; Mills, E.; Hashefi, M. Stimulation of human neutrophils by 5-oxo-6,8,11,14-eicosatetraenoic acid by a mechanism independent of the leukotriene B4 receptor. *J. Biol. Chem.* **1993**, *268*, 9280–9286. [CrossRef] [PubMed]
54. Libreros, S.; Shay, A.E.; Nshimiyimana, R.; Fichtner, D.; Martin, M.J.; Wourms, N.; Serhan, C.N. A New E-Series Resolvin: RvE4 Stereochemistry and Function in Efferocytosis of Inflammation-Resolution. *Front. Immunol.* **2020**, *11*, 631319. [CrossRef] [PubMed]
55. Perry, S.C.; Kalyanaraman, C.; Tourdot, B.E.; Conrad, W.S.; Akinkugbe, O.; Freedman, J.C.; Holinstat, M.; Jacobson, M.P.; Holman, T.R. 15-Lipoxygenase-1 biosynthesis of 7S,14S-diHDHA implicates 15-lipoxygenase-2 in biosynthesis of resolvin D5. *J. Lipid Res.* **2020**, *61*, 1087–1103. [CrossRef]
56. Chiang, N.; Fredman, G.; Backhed, F.; Oh, S.F.; Vickery, T.; Schmidt, B.A.; Serhan, C.N. Infection regulates pro-resolving mediators that lower antibiotic requirements. *Nature* **2012**, *484*, 524–528. [CrossRef] [PubMed]
57. De Marino, S.; Carino, A.; Masullo, D.; Finamore, C.; Marchiano, S.; Cipriani, S.; Di Leva, F.S.; Catalanotti, B.; Novellino, E.; Limongelli, V.; et al. Hyodeoxycholic acid derivatives as liver X receptor alpha and G-protein-coupled bile acid receptor agonists. *Sci. Rep.* **2017**, *7*, 43290. [CrossRef] [PubMed]
58. Farnegardh, M.; Bonn, T.; Sun, S.; Ljunggren, J.; Ahola, H.; Wilhelmsson, A.; Gustafsson, J.A.; Carlquist, M. The three-dimensional structure of the liver X receptor beta reveals a flexible ligand-binding pocket that can accommodate fundamentally different ligands. *J. Biol. Chem.* **2003**, *278*, 38821–38828. [CrossRef] [PubMed]
59. Krasowski, M.D.; Ni, A.; Hagey, L.R.; Ekins, S. Evolution of promiscuous nuclear hormone receptors: LXR, FXR, VDR, PXR, and CAR. *Mol. Cell Endocrinol.* **2011**, *334*, 39–48. [CrossRef] [PubMed]
60. Reschly, E.J.; Ai, N.; Welsh, W.J.; Ekins, S.; Hagey, L.R.; Krasowski, M.D. Ligand specificity and evolution of liver X receptors. *J. Steroid Biochem. Mol. Biol.* **2008**, *110*, 83–94. [CrossRef]
61. Heitel, P.; Achenbach, J.; Moser, D.; Proschak, E.; Merk, D. DrugBank screening revealed alitretinoin and bexarotene as liver X receptor modulators. *Bioorg. Med. Chem. Lett.* **2017**, *27*, 1193–1198. [CrossRef]
62. Landis, M.S.; Patel, H.V.; Capone, J.P. Oxysterol activators of liver X receptor and 9-cis-retinoic acid promote sequential steps in the synthesis and secretion of tumor necrosis factor-alpha from human monocytes. *J. Biol. Chem.* **2002**, *277*, 4713–4721. [CrossRef] [PubMed]

63. Lehmann, J.M.; Kliewer, S.A.; Moore, L.B.; Smith-Oliver, T.A.; Oliver, B.B.; Su, J.L.; Sundseth, S.S.; Winegar, D.A.; Blanchard, D.E.; Spencer, T.A.; et al. Activation of the nuclear receptor LXR by oxysterols defines a new hormone response pathway. *J. Biol. Chem.* **1997**, *272*, 3137–3140. [CrossRef] [PubMed]
64. Janowski, B.A.; Willy, P.J.; Devi, T.R.; Falck, J.R.; Mangelsdorf, D.J. An oxysterol signalling pathway mediated by the nuclear receptor LXR alpha. *Nature* **1996**, *383*, 728–731. [CrossRef] [PubMed]
65. Peet, D.J.; Janowski, B.A.; Mangelsdorf, D.J. The LXRs: A new class of oxysterol receptors. *Curr. Opin. Genet. Dev.* **1998**, *8*, 571–575. [CrossRef] [PubMed]
66. Janowski, B.A.; Grogan, M.J.; Jones, S.A.; Wisely, G.B.; Kliewer, S.A.; Corey, E.J.; Mangelsdorf, D.J. Structural requirements of ligands for the oxysterol liver X receptors LXRalpha and LXRbeta. *Proc. Natl. Acad. Sci. USA* **1999**, *96*, 266–271. [CrossRef] [PubMed]
67. Olkkonen, V.M.; Lehto, M. Oxysterols and oxysterol binding proteins: Role in lipid metabolism and atherosclerosis. *Ann. Med.* **2004**, *36*, 562–572. [CrossRef] [PubMed]
68. Rowe, A.H.; Argmann, C.A.; Edwards, J.Y.; Sawyez, C.G.; Morand, O.H.; Hegele, R.A.; Huff, M.W. Enhanced synthesis of the oxysterol 24(S),25-epoxycholesterol in macrophages by inhibitors of 2,3-oxidosqualene:lanosterol cyclase: A novel mechanism for the attenuation of foam cell formation. *Circ. Res.* **2003**, *93*, 717–725. [CrossRef] [PubMed]
69. Mol, M.J.; de Rijke, Y.B.; Demacker, P.N.; Stalenhoef, A.F. Plasma levels of lipid and cholesterol oxidation products and cytokines in diabetes mellitus and cigarette smoking: Effects of vitamin E treatment. *Atherosclerosis* **1997**, *129*, 169–176. [CrossRef]
70. Vejux, A.; Samadi, M.; Lizard, G. Contribution of cholesterol and oxysterols in the physiopathology of cataract: Implication for the development of pharmacological treatments. *J. Ophthalmol.* **2011**, *2011*, 471947. [CrossRef]
71. Wang, D.Q.; Afdhal, N.H. Good cholesterol, bad cholesterol: Role of oxysterols in biliary tract diseases. *Gastroenterology* **2001**, *121*, 216–218. [CrossRef]
72. Javitt, N.B.; Javitt, J.C. The retinal oxysterol pathway: A unifying hypothesis for the cause of age-related macular degeneration. *Curr. Opin. Ophthalmol.* **2009**, *20*, 151–157. [CrossRef] [PubMed]
73. Jeitner, T.M.; Voloshyna, I.; Reiss, A.B. Oxysterol derivatives of cholesterol in neurodegenerative disorders. *Curr. Med. Chem.* **2011**, *18*, 1515–1525. [CrossRef] [PubMed]
74. Muse, E.D.; Yu, S.; Edillor, C.R.; Tao, J.; Spann, N.J.; Troutman, T.D.; Seidman, J.S.; Henke, A.; Roland, J.T.; Ozeki, K.A.; et al. Cell-specific discrimination of desmosterol and desmosterol mimetics confers selective regulation of LXR and SREBP in macrophages. *Proc. Natl. Acad. Sci. USA* **2018**, *115*, E4680–E4689. [CrossRef]
75. Olkkonen, V.M.; Beaslas, O.; Nissila, E. Oxysterols and their cellular effectors. *Biomolecules* **2012**, *2*, 76–103. [CrossRef]
76. Rigamonti, E.; Chinetti-Gbaguidi, G.; Staels, B. Regulation of macrophage functions by PPAR-alpha, PPAR-gamma, and LXRs in mice and men. *Arterioscler. Thromb. Vasc. Biol.* **2008**, *28*, 1050–1059. [CrossRef]
77. Joseph, S.B.; Castrillo, A.; Laffitte, B.A.; Mangelsdorf, D.J.; Tontonoz, P. Reciprocal regulation of inflammation and lipid metabolism by liver X receptors. *Nat. Med.* **2003**, *9*, 213–219. [CrossRef]
78. Castrillo, A.; Joseph, S.B.; Marathe, C.; Mangelsdorf, D.J.; Tontonoz, P. Liver X receptor-dependent repression of matrix metalloproteinase-9 expression in macrophages. *J. Biol. Chem.* **2003**, *278*, 10443–10449. [CrossRef]
79. O'Brien, J.; Wilson, I.; Orton, T.; Pognan, F. Investigation of the Alamar Blue (resazurin) fluorescent dye for the assessment of mammalian cell cytotoxicity. *Eur. J. Biochem.* **2000**, *267*, 5421–5426. [CrossRef] [PubMed]
80. Rund, K.M.; Ostermann, A.I.; Kutzner, L.; Galano, J.M.; Oger, C.; Vigor, C.; Wecklein, S.; Seiwert, N.; Durand, T.; Schebb, N.H. Development of an LC-ESI(-)-MS/MS method for the simultaneous quantification of 35 isoprostanes and isofurans derived from the major n3- and n6-PUFAs. *Anal. Chim. Acta* **2018**, *1037*, 63–74. [CrossRef]
81. Kutzner, L.; Rund, K.M.; Ostermann, A.I.; Hartung, N.M.; Galano, J.M.; Balas, L.; Durand, T.; Balzer, M.S.; David, S.; Schebb, N.H. Development of an Optimized LC-MS Method for the Detection of Specialized Pro-Resolving Mediators in Biological Samples. *Front. Pharmacol.* **2019**, *10*, 169. [CrossRef]
82. Koch, E.; Mainka, M.; Dalle, C.; Ostermann, A.I.; Rund, K.M.; Kutzner, L.; Froehlich, L.F.; Bertrand-Michel, J.; Gladine, C.; Schebb, N.H. Stability of oxylipins during plasma generation and long-term storage. *Talanta* **2020**, *217*, 121074. [CrossRef] [PubMed]
83. Hartung, N.M.; Ostermann, A.I.; Immenschuh, S.; Schebb, N.H. Combined Targeted Proteomics and Oxylipin Metabolomics for Monitoring of the COX-2 Pathway. *Proteomics* **2021**, *21*, e1900058. [CrossRef] [PubMed]
84. Bornhorst, J.; Wehe, C.A.; Huwel, S.; Karst, U.; Galla, H.J.; Schwerdtle, T. Impact of manganese on and transfer across blood-brain and blood-cerebrospinal fluid barrier in vitro. *J. Biol. Chem.* **2012**, *287*, 17140–17151. [CrossRef]
85. Mallick, P.; Schirle, M.; Chen, S.S.; Flory, M.R.; Lee, H.; Martin, D.; Ranish, J.; Raught, B.; Schmitt, R.; Werner, T.; et al. Computational prediction of proteotypic peptides for quantitative proteomics. *Nat. Biotechnol.* **2007**, *25*, 125–131. [CrossRef] [PubMed]
86. Wilkins, M.R.; Gasteiger, E.; Bairoch, A.; Sanchez, J.C.; Williams, K.L.; Appel, R.D.; Hochstrasser, D.F. Protein identification and analysis tools in the ExPASy server. *Methods Mol. Biol.* **1999**, *112*, 531–552. [PubMed]
87. Gaudet, P.; Michel, P.A.; Zahn-Zabal, M.; Britan, A.; Cusin, I.; Domagalski, M.; Duek, P.D.; Gateau, A.; Gleizes, A.; Hinard, V.; et al. The neXtProt knowledgebase on human proteins: 2017 update. *Nucleic Acids Res.* **2017**, *45*, D177–D182. [CrossRef] [PubMed]
88. Johnson, M.; Zaretskaya, I.; Raytselis, Y.; Merezhuk, Y.; McGinnis, S.; Madden, T.L. NCBI BLAST: A better web interface. *Nucleic Acids Res.* **2008**, *36*, W5–W9. [CrossRef] [PubMed]

89. Fannes, T.; Vandermarliere, E.; Schietgat, L.; Degroeve, S.; Martens, L.; Ramon, J. Predicting tryptic cleavage from proteomics data using decision tree ensembles. *J. Proteome Res.* **2013**, *12*, 2253–2259. [CrossRef] [PubMed]
90. UniProt, C. UniProt: A worldwide hub of protein knowledge. *Nucleic Acids Res.* **2019**, *47*, D506–D515.
91. Hornbeck, P.V.; Zhang, B.; Murray, B.; Kornhauser, J.M.; Latham, V.; Skrzypek, E. PhosphoSitePlus, 2014: Mutations, PTMs and recalibrations. *Nucleic Acids Res.* **2015**, *43*, D512–D520. [CrossRef]
92. Hoofnagle, A.N.; Whiteaker, J.R.; Carr, S.A.; Kuhn, E.; Liu, T.; Massoni, S.A.; Thomas, S.N.; Townsend, R.R.; Zimmerman, L.J.; Boja, E.; et al. Recommendations for the Generation, Quantification, Storage, and Handling of Peptides Used for Mass Spectrometry-Based Assays. *Clin. Chem.* **2016**, *62*, 48–69. [CrossRef] [PubMed]
93. Lange, V.; Picotti, P.; Domon, B.; Aebersold, R. Selected reaction monitoring for quantitative proteomics: A tutorial. *Mol. Syst. Biol.* **2008**, *4*, 222. [CrossRef] [PubMed]
94. Gallien, S.; Duriez, E.; Domon, B. Selected reaction monitoring applied to proteomics. *J. Mass. Spectrom.* **2011**, *46*, 298–312. [CrossRef] [PubMed]
95. Krokhin, O.V.; Spicer, V. Peptide retention standards and hydrophobicity indexes in reversed-phase high-performance liquid chromatography of peptides. *Anal. Chem.* **2009**, *81*, 9522–9530. [CrossRef]
96. Kusebauch, U.; Campbell, D.S.; Deutsch, E.W.; Chu, C.S.; Spicer, D.A.; Brusniak, M.Y.; Slagel, J.; Sun, Z.; Stevens, J.; Grimes, B.; et al. Human SRMAtlas: A Resource of Targeted Assays to Quantify the Complete Human Proteome. *Cell* **2016**, *166*, 766–778. [CrossRef]

molecules

MDPI

Review

Naturally Occurring PCSK9 Inhibitors: An Updated Review

Jungmoo Huh [1] **and Hyunwoo Kim** [2,*]

[1] Research Institute of Pharmaceutical Sciences, College of Pharmacy, Seoul National University, Seoul 08826, Republic of Korea; goodhjm112@snu.ac.kr

[2] College of Pharmacy and Integrated Research Institute for Drug Development, Dongguk University-Seoul, Goyang 10326, Republic of Korea

* Correspondence: hwkim8906@dongguk.edu

Abstract: Proprotein convertase subtilisin/kexin type 9 (PCSK9) is a key modulator of low-density lipoprotein cholesterol (LDL-C) levels and emerged as an attractive therapeutic target for the treatment of hypercholesterolemia and cardiovascular diseases. Although statins and ezetimibe have been widely used to manage these disorders, concerns regarding side effects and high costs have driven ongoing efforts to search for alternative therapeutic candidates. To date, several classes of PCSK9 inhibitors, including monoclonal antibodies, oligonucleotides, proteins, and peptides, have been approved or are under clinical trials. In this review, we summarize 57 newly identified compounds derived from natural products showing inhibitory effects against PCSK9 reported between 2020 and April 2025. These compounds were isolated from 18 plants species and belong to various structural classes, including isoprenoids, flavonoids, alkaloids, and phenolic derivatives.

Keywords: PCSK9; natural products; LDL; hypercholesterolemia; cardiovascular disease

1. Introduction

Cholesterol plays a vital role as a key component of cell membranes in the human body, contributing to their structure, permeability, and fluidity [1]. In addition, cholesterol also acts as a precursor for the biosynthesis of steroid and sex hormones, bile acids, vitamin D, and lipoproteins [2,3]. Cholesterol is transported through the body in the form of lipoproteins, which circulate in the bloodstream [4]. These lipoproteins are classified into two types, low-density lipoproteins (LDLs) and high-density lipoproteins (HDLs) [2,4]. When cholesterol levels become imbalanced, particularly when they are elevated, the formation of plaque in blood vessels is facilitated, raising the risk of cardiovascular diseases, atherosclerosis, and hypercholesterolemia [2,4–6]. To reduce the risk of these diseases and manage elevated cholesterol levels, physicians have prescribed statin compounds and ezetimibe [2,7–9]. However, due to individual medical histories, genetic factors, and side effects of statin treatment, the need for alternative therapeutic agents has become increasingly important [10–14]. The proprotein convertase subtilisin/kexin type 9 (PCSK9) was discovered in 2003 [15], and it is mainly produced in the liver in humans [15]. PCSK9 plays a key role in cholesterol homeostasis by regulating the levels of LDL receptors (LDLRs) on hepatocyte surfaces [16,17]. PCSK9 binds to LDLRs and promotes their degradation in lysosomes, thereby reducing the ability to remove LDL cholesterol (LDL-C) from the bloodstream [16,18,19] (Figure 1). As a result, elevated PCSK9 activity leads to increased circulating LDL-C levels, contributing to a higher risk of atherosclerosis and cardiovascular disease [16,20]. Given its pivotal role in lipid metabolism,

PCSK9 has emerged as an attractive therapeutic target for managing hypercholesterolemia, liver diseases, and associated cardiovascular conditions.

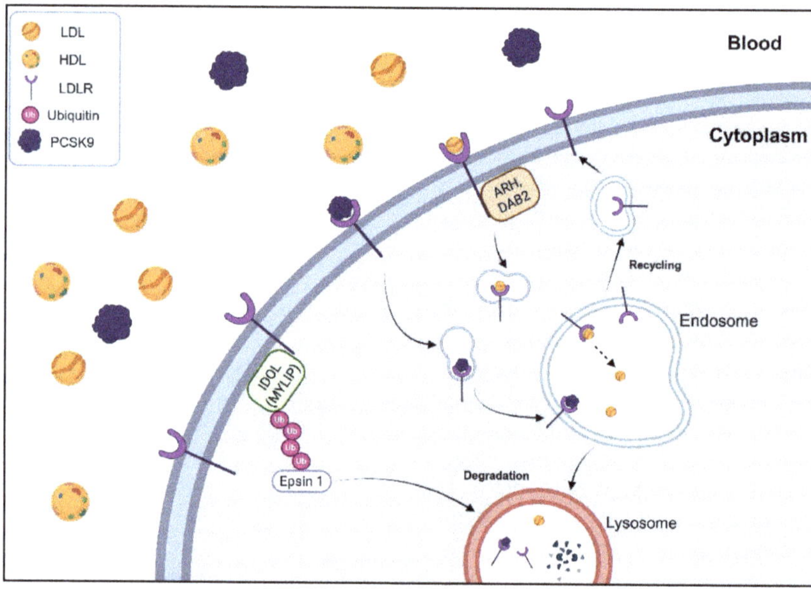

Figure 1. The role of PCSK9 in LDLR regulation. The figure was created with Biorender.com, with permission.

To date, several PCSK9 inhibitors have either been approved or are currently under clinical trials. Most of these inhibitors are monoclonal antibodies (alirocumab [21], evolocumab [22], bococizumab [23], LY3015014 [24], ongericimab/JS002 [25], tafolecimab/IBI306 [26], ebronucimab/AK102 [27], and recaticimab [28]), while others include proteins (LIB003 [29]), peptides (MK-0616 [30], NNC0385-0434 [31], and PCSK9 adnectin [32]), oligonucleotides (Inclisiran [33]), and small molecules (AZD0780 [34] and BMS-962476 [35]). Additionally, several natural-product-derived compounds such as berberine [16,17], ginkgolide B [36], lupin [17,36], polyphenols (quercetin [17], resveratrol [17,36], EGCG [17], and curcumin [36]), lycopene [17,36], etc., have been reported to exhibit inhibitory effects on PCSK9.

In this review, we summarize natural-product-derived compounds reported from 2020 to April 2025 that exhibit inhibitory effects on PCSK9 secretion. Although several reviews on PCSK9 inhibitors derived from natural products have been reported, they typically focused on specific disorders, such as atherosclerosis [37], hypercholesterolemia [36], and cardiovascular diseases [16,38,39], or reviewed the mechanisms associated with these diseases [16,36,40]. Some previous reviews on natural products have provided updates on well-known compounds such as berberine [16,17], resveratrol [17,41,42], and curcumin [17,42]. In contrast, this review compiles and categorizes a broad range of compounds that have been newly isolated from natural sources and confirmed to exhibit inhibitory effects against PCSK9.

2. Results and Discussion

A total of 350 compounds have been isolated from natural products and evaluated for their inhibitory activity against PCSK9 secretion from January 2020 to date. Among them are 57 newly identified compounds with PCSK9 inhibitory effects that were not covered in a previous review published in 2020 [17]. A diverse range of compound classes, including isoprenoids, flavonoids, alkaloids, and phenolic derivatives, were isolated from 18 different plant species.

2.1. The Naturally Occurring PCSK9 Inhibitors Reported in 2020

In 2020, Li et al. [43] isolated a cucurbitane-type triterpenoid derivative, 23,24-dihydrocucurbitacin B (**1**), from the ethanolic extract of *Trichosanthes cucumeroides* roots. Their work on HepG2 cells revealed that 23,24-dihydrocucurbitacin B (**1**) upregulated low-density lipoprotein receptor (LDLR) protein expression, leading to a dose-dependent elevation of 1,1′-dioctadecyl-3,3,3′,3′-tetramethylindocarbocyanine perchlorate-labeled low-density lipoprotein (DiI-LDL) uptake. Furthermore, the researchers investigated the relationship between PCSK9 and LDLRs, finding that 23,24-dihydrocucurbitacin B (**1**) reduced PCSK9 protein levels while increasing LDLR mRNA levels. Mechanistically, compound **1** modulates the transcription of PCSK9 and LDLRs via HNF1 and SRE1 motifs, respectively, by altering the nuclear levels of HNF-1α and SREBP2. These results were confirmed by dose-dependent Western blot and qPCR analyses using concentrations of 1, 2, 5, and 10 μM. The study also included in vivo experiments with HFD-fed hamsters, which showed that 23,24-dihydrocucurbitacin B (**1**) decreased total cholesterol, triglyceride, and LDL-C levels, while regulating LDLR and PCSK9 expression in the liver. A Western blot analysis of the liver tissues confirmed decreased PCSK9 and HNF-1α levels and increased LDLR and SREBP2 levels, along with the upregulation of SREBP2 target genes (HMGCR and HMGCS1).

A study by Zhang et al. [44] reported the isolation of 7 new and 20 known cucurbitacins from *T. cucumeroides*. To establish structure–activity relationships (SARs), the research team also synthesized 22 derivatives. A total of 47 cucurbitacins were then evaluated at 5 μM for their LDL uptake activity in HepG2 cells. The results showed that hexanorisocucurbitacin D (**2**) and isocucurbitacin D (**3**) exhibited greater LDL uptake than the positive control, nagilactone B. The subsequent SAR analysis demonstrated that a 2-oxo-3α-hydroxy A-ring is crucial for activity. It was also noted that while modifications at C-2, C-3, or C-16 had little impact, the presence of a carbonyl group at C-7 or a methoxy group at C-24 could enhance LDL uptake. Further Western blot analysis revealed that both hexanorisocucurbitacin D (**2**) and isocucurbitacin D (**3**) increased LDLR protein levels at concentrations of 5, 10, and 20 μM and decreased PCSK9 protein levels at 10 and 20 μM.

A study by Pel et al. [45] reported the purification of 31 compounds, including a stilbene dimer, flavonoids, and phenolic acids, from the methanolic extract of aerial parts of *Chromolaena odorata* to assess their inhibitory activity on PCSK9 expression in HepG2 cells. Prior to this isolation, the research team had already confirmed the inhibitory effects of the crude extract and its subsequent fractions. Among the isolated compounds, (+)-8b-*epi*-ampelopsin A (**4**), 5,6,7,4′-tetramethoxyflavanone (**5**), 5,6,7,3′,4′-pentamethoxyflavanone (**6**), acacetin (**7**), and uridine (**8**) showed inhibitory activity against PCSK9 mRNA expression, with IC_{50} values of 20.6, 21.4, 31.7, 15.0, and 13.7 μM, respectively.

Based on its potency and available quantity, 5,6,7,4′-tetramethoxyflavanone (**5**) was selected for deeper investigation. Further evaluation using Western blot analysis showed that treatment with 5,6,7,4′-tetramethoxyflavanone (**5**) at 10, 20, and 40 μM suppressed PCSK9 protein expression while elevating LDLR protein levels at 10 and 20 μM. These results suggest that the decrease in PCSK9 production led to enhanced LDLR protein recycling rather than its lysosomal degradation. Given that transcription factors like SREBP and HNF-1α are known to regulate PCSK9 [46], the authors investigated this pathway. Based on the observed suppression of HNF-1α mRNA expression, they concluded that

5,6,7,4′-tetramethoxyflavanone (**5**) might regulate PCSK9 via the HNF-1α pathway. The chemical structures of the PCSK9 inhibitors reported in 2020 are depicted in Figure 2, and the summarized information is presented in Table 1.

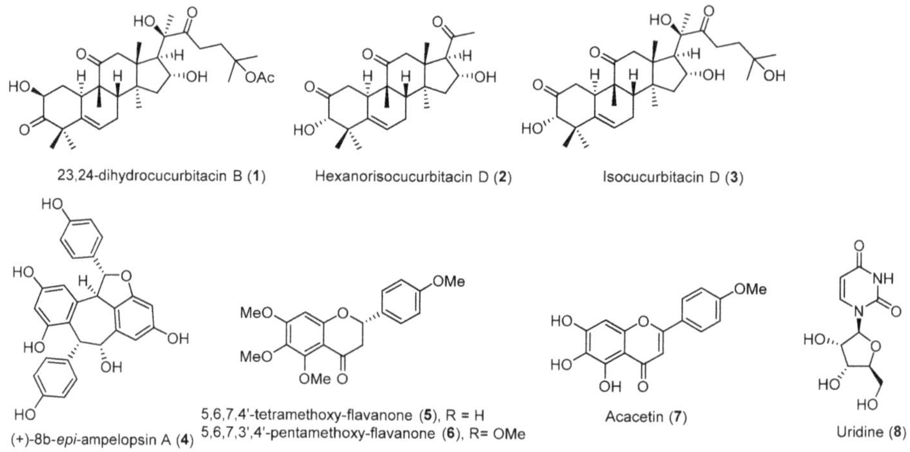

Figure 2. Chemical structures of the naturally occurring PCSK9 inhibitors reported in 2020.

Table 1. Summary of PCSK9 inhibitors reported in 2020.

No.	Compound Name	Compound Class	Origin/Source	Study Model	Activity Level	Mechanism	References
1	23,24-dihydrocucurbitacin B	Triterpenoid	*Trichosanthes cucumeroides* roots	HepG2 cells		Inhibits PCSK9 and HNF-1α level; increases LDLR and SREBP2 levels	[43]
				HFD-fed hamsters	30 mg/kg; 50% downregulated; 80% and 70% increased		
2	Hexanorisocucurtitacin D			HepG2 cells	5 μM; LDL uptake rate of 2.53	Suppresses PCSK9 mRNA; enhances LDLR mRNA	[44]
3	Isocucurbitacin D				5 μM; LDL uptake rate of 2.47		[44]
4	(+)-8b-*epi*-ampelopsin A	Stilbene	*Chromolaena odorata* aerial parts	HepG2 cells	IC_{50} 20.6 μM	Inhibits PCSK9 mRNA expression	[45]
5	5,6,7,4′-tetramethoxyflavanone	Flavonoid			IC_{50} 21.4 μM		
6	5,6,7,3′,4′-pentamethoxyflavanone				IC_{50} 31.7 μM		
7	Acacetin				IC_{50} 15.0 μM		
8	Uridine	Nucleic acid			IC_{50} 13.7 μM		

2.2. *The Naturally Occurring PCSK9 Inhibitors Reported in 2021*

In 2021, a study by Nhoek et al. [47] described the isolation of 7 new sesquiterpenes and 12 known compounds from the aerial parts of *Salvia plebeian*. All isolated compounds were tested at 50 μM for inhibitory effects on PCSK9 mRNA expression in HepG2 cells. Among them, plebeic acid A (**9**), (1*S*,5*S*,8*S*,10*R*)-1-acetoxy-8-methoxy-2-oxoeudesman-3,7(11)-dien-8,12-olide (**10**), and eudebeiolide B (**11**) showed significant inhibitory activity, with IC_{50} values of 24.4, 25.2, and 27.8 μM, respectively. The two most potent compounds, plebeic acid A (**9**) and eudebeiolide B (**11**), also moderately upregulated LDLR mRNA expression. A subsequent Western blot analysis revealed that both compounds slightly reduced PCSK9 protein and increased LDLR protein levels at

10 and 50 μM, though the effects were less potent than those of the positive control, berberine. The investigation also included a PCSK9-LDLR binding assay, but no isolates showed activity.

Also in 2021, Weng et al. [48] reported 20 dammarane-type triterpenoidal saponins from *Gynostema pentaphyllum*. To assess lipid-lowering activity, the authors selected eight compounds for PCSK9 inhibition experiments in HepG2 cells. The results showed that the gypenosides LXXXIX (**12**), XC (**13**), and XCI (**14**), and the ginsenoside Rg5 (**15**) significantly inhibited simvastatin-induced PCSK9 expression at 20 μM. The gypenosides LXXXIX (**12**) and XC (**13**) and the ginsenoside R_{g5} (**15**), in particular, demonstrated potent suppression of PCSK9 expression even at a lower concentration of 10 μM.

Kim et al. [49] isolated and reported 4 new prenylated flavonoid glycosides along with 18 known compounds from the dried aerial parts of *Epimedium koreanum*. They tested all isolated compounds and their inhibitory effects of PCSK9 mRNA expression and modulation of LDLR mRNA expression in HepG2 cells. Ten compounds, icariside I (**16**), ikarisoside A (**17**), icariin (**18**), anhydroicaritin 3-*O*-β-D-fucopyranosyl(1→2)-rhamnopyranoside-7-*O*-β-D-glucoside (**19**), korepimedoside A (**20**), epimedokoreanoside I (**21**), korepimeoside C (**22**), epimedin L (**23**), caohuoside B (**24**), and epimedoicarisoside A (**25**), showed inhibitory effects against PCSK9 expression at 10 μM. Notably, only ikarisoside A (**17**) elevated LDLR mRNA expression, suggesting that this compounds has potential for use as a cholesterol-lowering drug.

Woo et al. [50] reported five new selaginellin derivatives along with one known selaginellin from *Selaginella tamariscina* roots and rhizophores. Among the isolated compounds, selaginpulvilin U (**26**) showed the highest upregulation of LDLR-related genes at 50 μM in human HepG2 cells. The authors further evaluated selaginpulvilin U (**26**) for its effects on LDLR transcript and protein levels in a dose-dependent manner and found that LDLR expression was regulated via SREBPs.

Ahn et al. [51] isolated two lignan dimers, obovatalins A (**27**) and B (**28**), along with magnolol (**29**) from the dried bark of *Magnolia obovate*. The authors investigated the effect of these compounds on PCSK9 expression levels, and the results showed that compounds **27**–**29** exhibited inhibitory effects on PCSK9 expression in HepG2 cells, with IC_{50} values of 12.0, 45.4, and 22.9 μM, respectively. In a further investigation, obovatalin A (**27**) notably reduced PCSK9 protein levels and increased LDLR expression. The chemical structures of the PCSK9 inhibitors reported in 2021 are shown in Figure 3, and the summarized information is presented in Table 2.

Table 2. Summary of PCSK9 inhibitors reported in 2021.

No.	Compound Name	Compound Class	Origin/Source	Study Model	Activity Level	Mechanism	References
9	Plebeic acid A	Sesquiterpene	*Salvia plebeian* roots	HepG2 cells	IC_{50} 24.4 μM	Inhibitory effects on PCSK9 mRNA expression; upregulates LDLR mRNA expression	[47]
10	(1*S*,5*S*,8*S*,10*R*)-1-acetoxy-8-methoxy-2-oxoeudesman-3,7(11)-dien-8,12-olide				IC_{50} 25.2 μM		
11	Eudebeiolide B				IC_{50} 27.8 μM		
12	Gypenoside LXXXIX	Triterpenoidal saponin	Whole herb of *Gynostema pentaphyllum*	HepG2 cells	20 μM	Inhibition against simvastatin-induced PCSK9 expression	[48]
13	Gypenoside XC				10 and 20 μM		
14	Gypenoside XCI				10 and 20 μM		
15	Ginsenoside Rg5				10 and 20 μM		

Table 2. *Cont.*

No.	Compound Name	Compound Class	Origin/Source	Study Model	Activity Level	Mechanism	References
16	Icariside I	Flavonoid glycoside	*Epimedium koreanum* aerial parts	HepG2 cells	10 μM	Inhibits PCSK9 mRNA expression	[49]
17	Ikarisoside A					Inhibits PCSK9 mRNA expression; increases LDLR mRNA expression	
18	Icariin					Inhibits PCSK9 mRNA expression	
19	Anhydroicaritin 3-*O*-β-D-fucopyranosyl(1→2)-rhamnopyranoside-7-*O*-β-D-glucoside						
20	Korepimedoside A						
21	Epimedokoreanoside I						
22	Korepimeoside C						
23	Epimedin L						
24	Caohuoside B						
25	Epimedoicarisoside A						
26	Selaginpulvilin U	Selaginellin derivative	*Selaginella tamariscina* roots and rhizophores	HepG2 cells	50 μM	Increases LDLR expression	[50]
27	Obovatalin A	Lignan	Dried bark of *Magnolia obovate*	HepG2 cells	IC_{50} 12.0 μM	Inhibitory effects on PCSK9 protein levels and increases LDLR expression	[51]
28	Obovatalin B				IC_{50} 45.4 μM		
29	Magnolol				IC_{50} 22.9 μM		

2.3. *The Naturally Occurring PCSK9 Inhibitors Reported in 2022*

In 2022, Huang et al. [52] isolated and reported dammarane-type saponins, gypenosides LXXXXI–LXXXVII, together with four known compounds from *G. pentaphyllum*. All isolated compounds were assessed for their effect on PCSK9 expression in HepG2 cells. PCSK9 expression was measured by ELISA in LPDS-induced HepG2 cells, and cell viability was assessed via MTT assay to exclude cytotoxic compounds. Several compounds that did not show cytotoxicity at 20 μM were selected for further analysis. Gypenosides LXXXII (**30**), LXXXV (**31**), and LXXXVII (**32**) showed PCSK9 inhibitory effects at 10 μM, although LXXXVII (**32**) showed some cytotoxicity. Notably, gypenoside LXXXII (**32**) also exhibited inhibitory activity at 5 μM. Further SAR analysis showed that the side chain at C-17 with the double bond at C-24 and C-25, the hydroxyl group at C-12, the oligosugar at C-20, and the methyl group at C-10 might be essential for inhibitory effects against PCSK9.

Zhang et al. [53] reported 40 compounds, including 6 new triterpenoids, alisolinal A–F, from the rhizome of *Alisma plantago-aquatica* in 2022, evaluating the promoted LDL uptake of all isolates in HepG2 cells using the DiI-LDL uptake quantified assay. Among the tested compounds, 17 compounds exhibited significant LDL-uptake-promoting activities, with 9 protostane-type triterpenoids showing strong activity. Among them, alisol A 23-acetate (**33**), alisol A 24-acetate (**34**), 16-oxo-11-anhydroalisol A (**35**), and alisol B 23-acetate (**36**) showed the most potent activity. Zhang et al. also reported the SAR analysis of 40 compounds in relation to LDL uptake. The C-17 spirost protostane-type triterpenoids with the *S* configuration showed higher activity than those with the *R* configuration. In addition, dihydroxylation of C-25 alisol A-type triterpenes and esterification of the hydroxyl group at C-24 in alisol F-type compounds led to enhanced activity. The isolated compounds were also assessed for their inhibitory effects on PCSK9 expression in HepG2 cells. Among them, alisol G (**37**) and alisolinal C (**38**) showed significant inhibition rates of 46% and 58%, respectively. Furthermore, alisol G (**37**) also showed a 55.2% inhibition of PCSK9 protein expression in Western blot analysis. The authors carried out further investigations with alisol G (**37**) and found that at 10 μM it decreased PCSK9 mRNA expression, increased LDLR mRNA expression, and promoted LDL uptake, whereas at 1 μM it showed no significant activity.

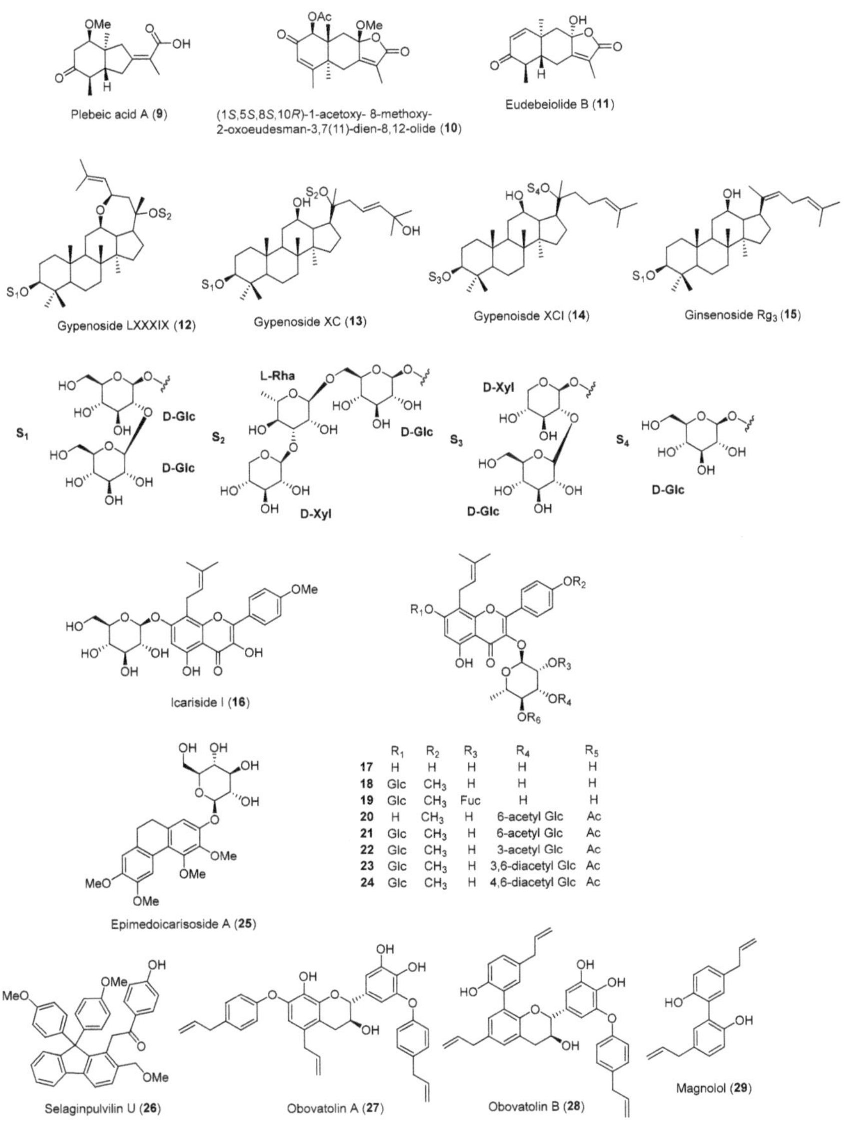

	R_1	R_2	R_3	R_4	R_5
17	H	H	H	H	H
18	Glc	CH_3	H	H	H
19	Glc	CH_3	Fuc	H	H
20	H	CH_3	H	6-acetyl Glc	Ac
21	Glc	CH_3	H	6-acetyl Glc	Ac
22	Glc	CH_3	H	3-acetyl Glc	Ac
23	Glc	CH_3	H	3,6-diacetyl Glc	Ac
24	Glc	CH_3	H	4,6-diacetyl Glc	Ac

Figure 3. Chemical structures of the naturally occurring PCSK9 inhibitors reported in 2021.

Huh et al. [54] isolated and reported acylated saponins, flavonoid glycosides, and (+)-catechin (**39**) from the fruits of *Stewartia koreana*. The authors screened the isolated compounds for their regulatory activity on PCSK9 and LDLR expression, but only (+)-catechin (**39**) exhibited inhibitory effects on PCSK9 mRNA levels at 50 μM without cytotoxicity.

Consequently, Huh et al. [54] conducted Western blot analysis with (+)-catechin (**39**), which showed upregulation of LDLR mRNA levels and downregulation of PCSK9 mRNA levels.

Pel et al. [55] reported 22 compounds, including 2 new isocoumarins and a new benzofuran, from the dried roots of *Lysimachia vulgaris*. All isolated compounds were tested for their inhibitory effects on PCSK9 and LDLR mRNA expression. Among them, 8′*Z*,11′*Z*-octadecadienyl-6,8-dihydroxyisocoumarin (**40**) and 5-*O*-methylembelin (**41**) inhibited PCSK9 mRNA expression at 20 μM significantly. The authors further evaluated 8′*Z*,11′*Z*-octadecadienyl-6,8-dihydroxyisocoumarin (**40**) and 5-*O*-methylembelin (**41**) at various concentrations, and the IC_{50} values for PCSK9 mRNA inhibition were 11.9 and 4.9 μM, respectively. Pel et al. also examined the effects of these two compounds on LDLR-related genes, and the results revealed that PCSK9 mRNA expression was downregulated by SREBP2.

Pel et al. [56] published another paper in 2022, in which they isolated 31 compounds from the roots and rhizomes of *Sophora tonkinensis*. Of these isolated compounds, (+)-isolariciresinol (**42**) showed suppressive effects on PCSK9 protein expression and decreased LDLR protein levels at 10 and 50 μM, as observed in Western blot analysis. Furthermore, (+)-isolariciresinol (**42**) downregulated HNF1α and SREBP mRNA expression, leading to reduced expression of both PCSK9 and LDLR proteins. The chemical structures of the PCSK9 inhibitors reported in 2022 are shown in Figure 4, and the summarized information is presented in Table 3.

Table 3. Summary of PCSK9 inhibitors reported in 2022.

No.	Compound Name	Compound Class	Origin/Source	Study Model	Activity Level	Mechanism	References
30	Gypenoside LXXXII	Triterpenoidal saponin	Whole herb of *Gynostema pentaphyllum*	HepG2 cells	5, 10 and 20 μM	Inhibition against LPDS-induced PCSK9 expression	[52]
31	Gypenoside LXXXV				10 and 20 μM		
32	Gypenoside LXXXVII				20 μM		
33	Alisol A 23-acetate	Triterpene	*Alisma plantago-aquatica* rhizomes	HepG2 cells	10 μM	Inhibits PCSK9 mRNA expression	[53]
34	Alisol A 24-acetate	Triterpene					
35	16-oxo-11-anhydroalisol A	Triterpene					
36	Alisol B 23-acetate	Triterpene					
37	Alisol G	Triterpene			58%, 10 μM		
38	Alisolinal C	Triterpene			46%, 10 μM		
39	(+)-Catechin	Flavonoid	*Stewartia koreana* fruits	HepG2 cells	50 μM	Suppresses PCSK9 protein levels and increases LDLR levels	[54]
40	8′*Z*,11′*Z*-octadecadienyl-6,8-dihydroxy-isocoumarin	Isocoumarin	*Lysimachia vulgaris* roots	HepG2 cells	IC_{50} 11.9 μM	Inhibits PCSK9 mRNA expression	[55]
41	5-*O*-methylembelin	Benzofuran			IC_{50} 4.9 μM		

Table 3. *Cont.*

No.	Compound Name	Compound Class	Origin/Source	Study Model	Activity Level	Mechanism	References
42	(+)-Isolariciresinol	Lignan	*Sophora tonkinensis* rhizomes	HepG2 cells	10 and 50 μM	Downregulates HNF1α and SREBP mRNA expression; reduces expression of PCSK9 and LDLR protein	[56]

LPDS: lipoprotein-deficient serum.

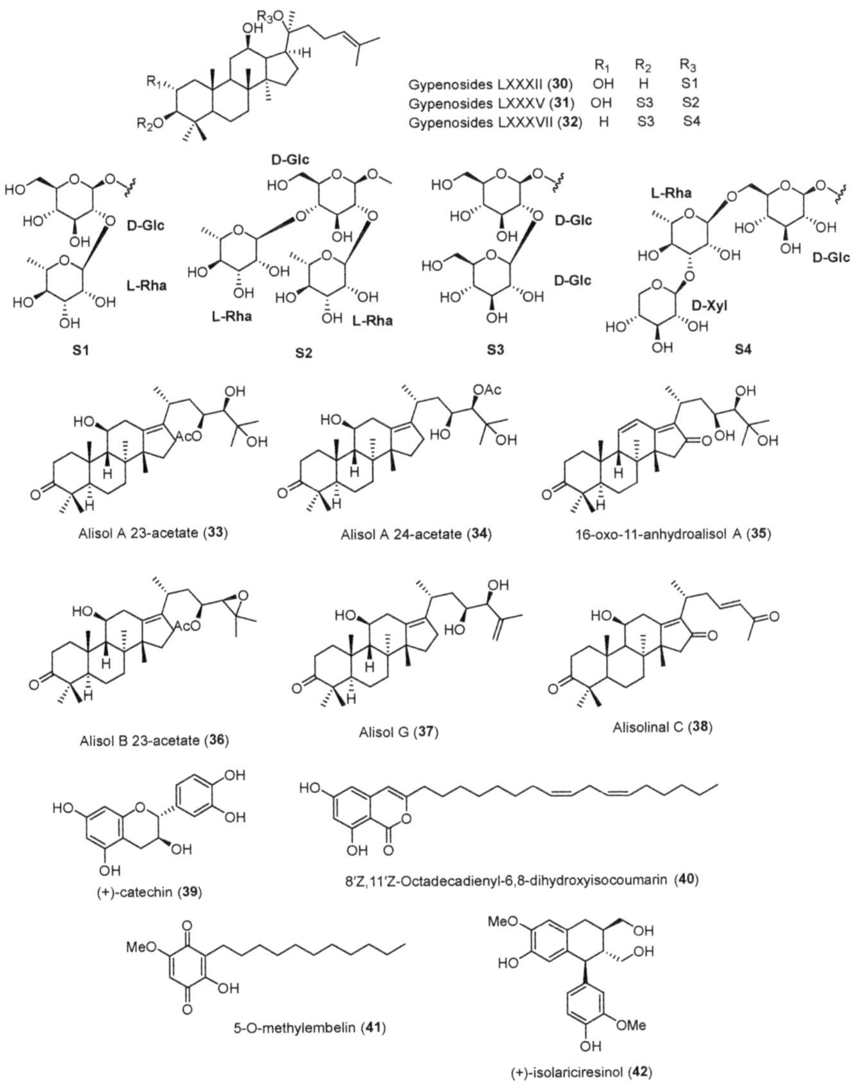

Figure 4. Chemical structures of the naturally occurring PCSK9 inhibitors reported in 2022.

2.4. The Naturally Occurring PCSK9 Inhibitors Reported in 2023 and 2024

Nhoek [57] isolated and reported 14 new clerodane diterpenoids from the fruits of *Casearia grewiifolia* in 2023. The isolated compounds were evaluated for their regulatory effects on LDLRs, PCSK9, and IDOL expression in HepG2 cells at 20 μM. LDLRs are known to be degraded by PCSK9 and IDOL [58,59]. Among the isolates, grewiifolin C (**43**) showed the strongest inhibition of PCSK9 and IDOL mRNA expression and was chosen for further immunoblot analysis. However, grewiifolin C (**43**) did not exhibit notable inhibitory activity against PCSK9 or IDOL protein expression at 20 and 40 μM.

An et al. [60] isolated 17 compounds, including 3 new acyclic triterpenoids, from the dried seeds of *Alpinia katsumadai*. Among these acyclic triterpenoids, both (3*R*,20*S*)-2,3,20-Trihydroxy-2,6,10,15,19,23-hexamethyl-tetracosa-6,10,14,18,22-pentaene (**44**) and (3*R*,5*S*)-2,3,5-Trihydroxy-2,6,10,15,19,23-hexamethyl-tetracosa-6,10,14,18,22-pentaene (**45**) showed significant inhibitory effects on PCSK9 secretion at 10 and 20 μM, respectively. In addition, compounds **44** and **45** were further evaluated for their effects on PCSK9 and LDLR mRNA expressions. The results showed that compounds **44** and **45** markedly suppressed PCSK9 mRNA levels and promoted LDLR mRNA expression. The authors also tested these two compounds using various concentrations ranging from 0.625 to 10 μM and 1.25 to 20 μM, respectively, and the IC_{50} values were obtained as 2.94 μM for **44** and 15.08 μM for **45**.

In 2024, Hu et al. [61] purified 12 compounds, including 9 amide alkaloids and 3 neolignans, from the aerial part of *Piper hongkongense* and assessed their PCSK9 inhibitory activities in HepG2 cells using the PCSK9 AlphaLISA screening method. In the PCSK9 AlphaLISA screening, hongkongensine C (**46**) and kadsurenone (**47**) showed 38.4% and 52.0% inhibition rates at 5 μM, respectively, and the positive control, berberine, exhibited an inhibition rate of 55.6% at 5 μM. In addition, kadsurenone (**47**) demonstrated dose-dependent inhibition of PCSK9 protein levels in HepG2 cells.

Son et al. [62] reported 5 new compounds, along with 27 known compounds, isolated from the roots of *Cynanchum wilfordii*, and evaluated their inhibitory effects on PCSK9 secretion. The authors conducted a water-soluble tetrazolium-8 (WST-8) assay with 10 μM to assess the cytotoxicity of the isolated compounds in HepG2 cells, and none of the compounds exhibited cytotoxicity. The isolates were then tested for their inhibitory effects on PCSK9 secretion using ELISA screening, in which several compounds showed moderate activity. Among the tested compounds, betulinic acid and (3*S*,8*S*,9*S*,10*R*,13*S*,14*S*,17*S*,22*R*)-24-methylcholesta-5,20,24-trien-3,22-ol (**48**) demonstrated strong inhibitory effects. Since the PCSK9-inhibitory activity of betulinic acid had already been reported [63], the authors selected (3*S*,8*S*,9*S*,10*R*,13*S*,14*S*,17*S*,22*R*)-24-methylcholesta-5,20,24-trien-3,22-ol (**48**) for further investigation. The compound was evaluated for its LDLR and PCSK9 mRNA expression, and the results showed suppressed PCSK9 expression with 2.5 (63%), 5 (27%), and 10 (27%) μM and slightly increased but no significant LDLR mRNA expression. The authors also conducted a Western blot analysis, and the PCSK9 protein levels with this compound were remarkably decreased at 5 and 10 μM. To assess the effect of **48** on transcription factors, qPCR experiments were conducted. Previous studies have identified SREBP1/2 and HNF1α as major regulators of PCSK9 expression [64], and berberine, a known PCSK9 inhibitor, has been reported to down-regulate both factors [65]. However, unlike berberine, compound **48** was found to increase the mRNA levels of SREBP 1/2 and HNF1α, suggesting that it regulates PCSK9 via a different pathway. Further analysis revealed that the compound also upregulated the expression of the forkhead box protein O1 (FOXO1) and FOXO3, alternative transcriptional regulators of PCSK9 [66]. In particular, FOXO3 was shown to bind to the PCSK9 promoter and interact with SIRT6, thereby suppressing PCSK9 gene expression through histone H3 deacetylation. Moreover, FOXO3 competitively inhibited HNF1α-mediated upregulation by binding to the

HNF1α-binding site on the PCSK9 promoter. These findings indicate that the compound downregulates PCSK9 by increasing FOXO3 levels. Based on these results, Son et al. [62] evaluated the effect of co-treatment with atorvastatin and compound **48**. While administration of 10 μM atorvastatin alone significantly increased PCSK9 mRNA expression, co-treatment with 2.5 μM of (3*S*,8*S*,9*S*,10*R*,13*S*,14*S*,17*S*,22*R*)-24-methylcholesta-5,20,24-trien-3,22-ol (**48**) markedly reduced PCSK9 mRNA levels compared to atorvastatin alone.

In 2024, Wei et al. [67] isolated and reported four new isoquinoline alkaloids from *Hypecoum erectum* and evaluated their PCSK9 inhibition effects. All four isolated compounds and the positive control, PF-06446846, were evaluated for their binding affinity with PCSK9 by surface plasmon resonance (SPR) analysis. Among the tested compounds, hypecotumines C (**51**) and D (**52**) showed moderate affinity strength at 95.1 and 59.9 μM, respectively, against PCSK9, compared to hypecotumines A (**49**) and B (**50**). These results suggest that the methylenedioxy moieties located at C-3′ and C-4′ contribute significantly to the affinity strength. In addition, the authors carried out a Western blot assay to assess the protein levels of PCSK9 and LDLRs, and all isolates showed a decreased expression of PCSK9 and an increased expression of LDLR mRNA. Furthermore, all compounds were tested to identify the binding sites on the PCSK9 protein through molecular docking (PDB ID: 6U3X), and the results exhibited that hypecotumine D (**52**) showed π–cation interactions with ARG-458 and formed a salt bridge with ASP-360, resulting in a binding pattern similar to that of PF-06446846. These interactions were not observed in hypecotumines A–C (**49**–**51**), which may explain the better affinity of hypecotumine D (**52**).

Lee et al. [68] confirmed 20 compounds from the whole plants of *Jacobaea vulgaris*. Among the isolated compounds, two stilbene derivatives, 3′-dehydroxy gancaonin R (**53**) and gancaonin R 3-acetate (**54**), significantly inhibited PCSK9 mRNA expression and promoted LDLR mRNA expression at 20 μM. The authors also tested these two compounds (**53** and **54**) at various concentrations ranging from 6.25 to 50 μM and determined their IC_{50} values for PCSK9 protein inhibition to be 16.1 and 20.6 μM, respectively. In addition, the effects of the compounds on LDLRs, PCSK9, and IDOL protein levels were examined, and the results showed that 3′-dehydroxy gancaonin R (**53**) upregulated the mature form of LDLR protein, while gancaonin R 3-acetate (**54**) increased LDLR protein levels and suppressed IDOL protein expression. The chemical structures of the PCSK9 inhibitors reported in 2023 and 2024 are depicted in Figure 5, and the summarized information is presented in Table 4.

Table 4. Summary of PCSK9 inhibitors reported in 2023 and 2024.

No.	Compound Name	Compound Class	Origin/Source	Study Model	Activity Level	Mechanism	References
43	Grewiifolin C	Diterpene	*Casearia grewiifolia* fruits	HepG2 cells	20 μM	Inhibits PCSK9 and IDOL mRNA expression	[57]
44	(3*R*,20*S*)-2,3,20-trihydroxy-2,6,10,15,19,23-hexamethyl-tetracosa-6,10,14,18,22-pentaene	Acyclic triterpenoid	Dried seeds of *Alpinia katsumadai*	HepG2 cells	IC_{50} 2.94 μM	Inhibition of PCSK9 mRNA expression	[60]
45	(3*R*,5*S*)-2,3,5-trihydroxy-2,6,10,15,19,23-hexamethyl-tetracosa-6,10,14,18,22-pentaene				IC_{50} 15.08 μM		
46	Hongkongensine C	Amide alkaloid	Aerial part of *Piper hongkongense*	HepG2 cells	5 μM, 38.4%	Inhibitory activity against PCSK9 expression	[61]
47	Kadsurenone	Lignan			5 μM, 52.0%		

Table 4. *Cont.*

No.	Compound Name	Compound Class	Origin/Source	Study Model	Activity Level	Mechanism	References
48	(3*S*,8*S*,9*S*,10*R*,13*S*,14*S*,17*S*,22*R*)-24-methylcholesta-5,20,24-trien-3,22-ol	Triterpene	*Cynanchum wilfordii* roots	HepG2 cells	2.5 (63%), 5 (27%), and 10 (27%) μM	Suppresses PCSK9 expression	[63]
49	Hypecotumine A	Isoquinoline alkaloids	Whole herb of *Hypecoum erectum*	Affinity with PCSK9 protein by SPR analysis	K_D 306.0 μM	Downregulates PCSK9 protein levels; upregulates LDLR protein levels	[67]
50	Hypecotumine B				K_D 248.0 μM		
51	Hypecotumine C				K_D 95.1 μM		
52	Hypecotumine D				K_D 59.9 μM		
53	3′-dehydroxy gancaonin R	Stilbenes	Whole herb of *Jacobaea vulgaris*	HepG2 cells	IC_{50} 16.1 μM	Inhibition of PCSK9 mRNA expression; upregulation of LDLR protein levels	[68]
54	Gancaonin R 3-acetate				IC_{50} 20.6 μM		

SPR: surface plasmon resonance.

2.5. The Naturally Occurring PCSK9 Inhibitors Reported in 2025

An et al. [69] confirmed a total of 16 cycloartane-type triterpenoids, including 9 new compounds, from the ethanolic extract of *Combretum quadrangulare* twigs and evaluated the isolates for PCSK9 secretion inhibitory activities. All isolated compounds were tested for their inhibitory activity against PCSK9 protein secretion at 20 μM. Among them, combretanol A (**55**), combretanone H (**56**), and combretic acid A (**57**) showed significant inhibition. In addition, a qPCR assay was also conducted using berberine as a positive control alongside the three compounds, revealing suppressed PCSK9 mRNA expression and promoted LDLR mRNA expression. In Western blot analysis, all three compounds reduced PCSK9 protein levels and downregulated mature PCSK9 (65 kDa), while only combretic acid A (**57**) significantly increased LDLR protein levels. Furthermore, to assess their ability to counteract atorvastatin-induced PCSK9 elevation, the three compounds (**55–57**) were co-treated with atorvastatin. All three compounds demonstrated inhibitory effects on atorvastatin-induced PCSK9 expression at 10 μM. Of these, combretic acid A (**57**) was selected for further evaluation due to its ability to increase LDLR protein expression. DiI-LDL staining in HepG2 cells confirmed that combretic acid A (**57**) remarkably increased LDL uptake at 10 μM. A pharmacokinetic study of combretic acid A (**57**) was performed in mice via intraperitoneal (IP) injection. It was rapidly absorbed and extensively distributed in the liver, where its concentration remained notably higher than in plasma from 60 to 300 min (T/P ratios > 1). Additionally, the $AUC_{0\text{-}300}$ of combretic acid A (**57**) in the liver (60.8 μg min/g tissue) was notably higher than that in plasma AUC (7.62 μg min/mL), indicating that the compound remains and acts primarily in the liver. Given that PCSK9 is predominantly produced and secreted in the liver, these findings suggest that combretic acid A (**57**) has potential as a promising therapeutic candidate for lowering PCSK9 production. The chemical structures of the PCSK9 inhibitors reported in 2025 are shown in Figure 6, and the summarized information is presented in Table 5.

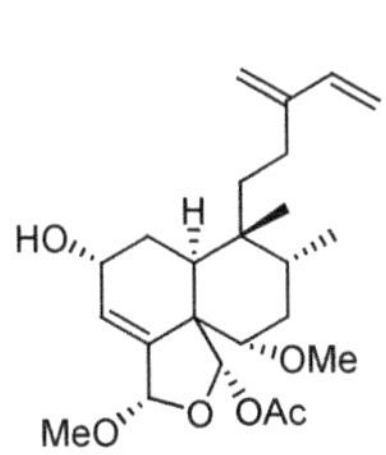

Grewiifolin C (**43**)

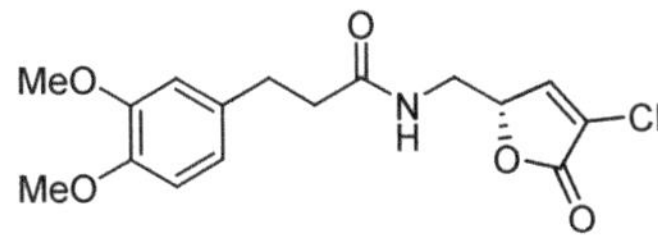

Hongkongensine C (**46**)

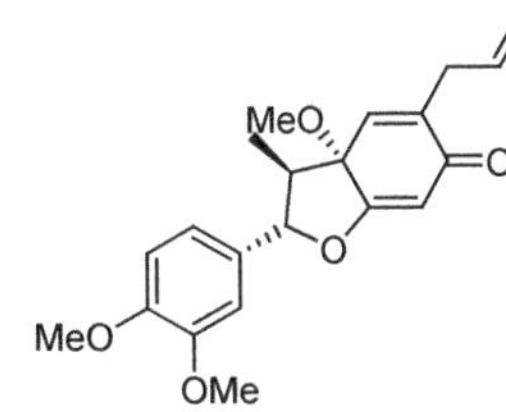

Kadsurenone (**47**)

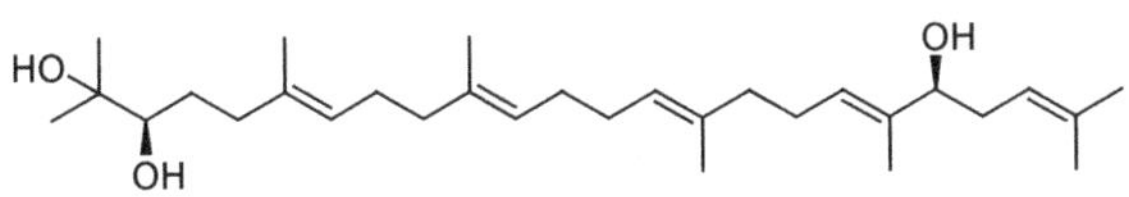

(3*R*,20*S*)-2,3,20-trihydroxy-2,6,10,15,19,23-hexamethyl-tetracosa-6,10,14,18,22-pentaene (**44**)

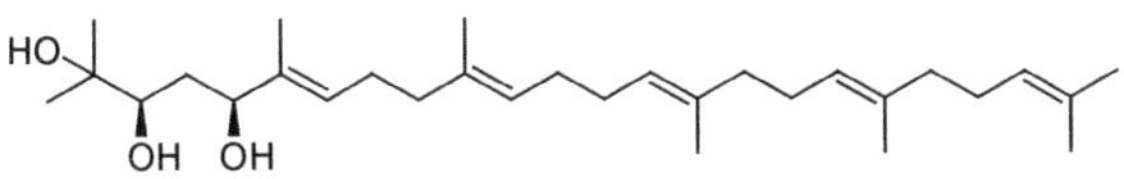

(3*R*,5*S*)-2,3,5-trihydroxy-2,6,10,15,19,23-hexamethyl-tetracosa-6,10,14,18,22-pentaene (**45**)

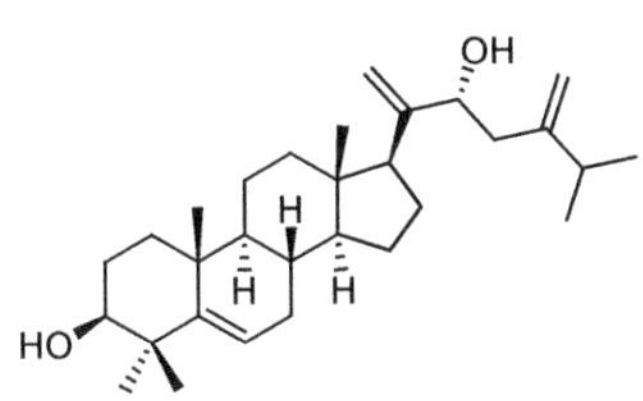

(3S,8S,9S,10R,13S,14S,17S,22R)-24-methylcholesta-5,20,24-trien-3,22-ol (**48**)

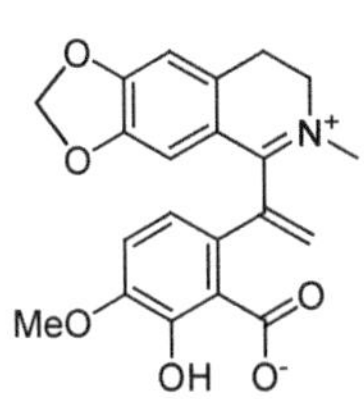

Hypecotumin A (**49**)

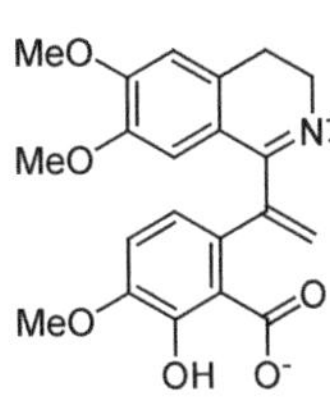

Hypecotumin B (**50**)

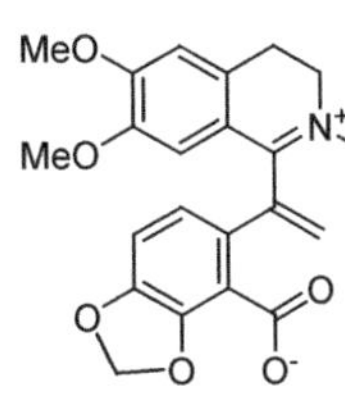

Hypecotumin C (**51**)

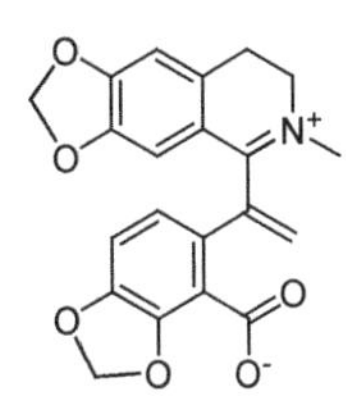

Hypecotumin D (**52**)

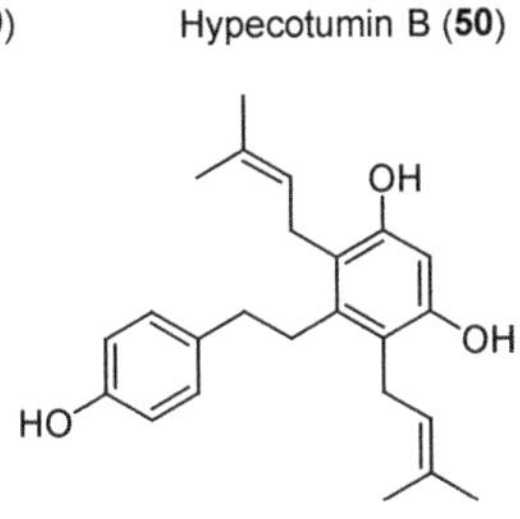

3′-Dehydroxy gancaonin R (**53**)

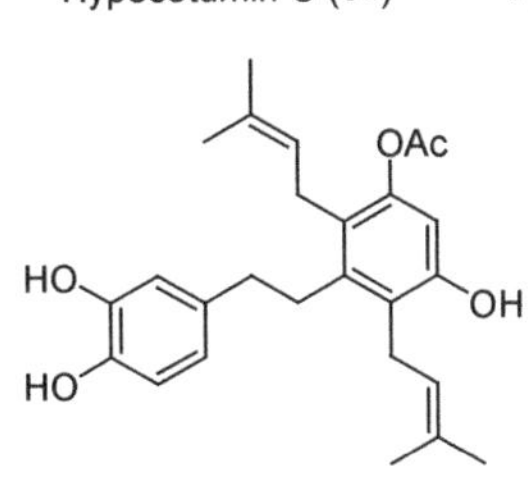

Gancaonin R 3-acetate (**54**)

Figure 5. Chemical structures of the naturally occurring PCSK9 inhibitors reported in 2023 and 2024.

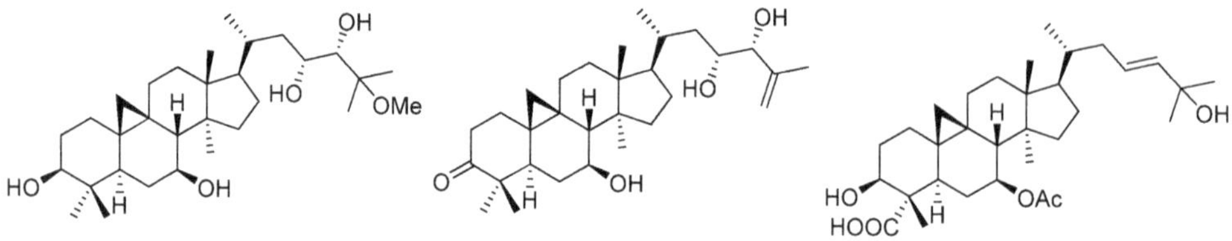

combretanol A (**55**) combretanone H (**56**) combretic acid A (**57**)

Figure 6. Chemical structures of the naturally occurring PCSK9 inhibitors reported in 2025.

Table 5. Summary of PCSK9 inhibitors reported in 2025.

No.	Compound Name	Compound Class	Origin/Source	Study Model	Activity Level	Mechanism	References
55	Combretanol A	Triterpenoid	*Combretum quadrangulare* twigs	HepG2 cells	20 μM	Suppresses PCSK9 mRNA expression; promotes LDLR mRNA expression	[69]
56	Combretanone H				5 and 10 μM		
57	Combretic acid A				2.5, 5, 10, and 20 μM		

3. Methodology

The keywords used in the review were "PCSK9 inhibitors", "Natural", and "Plant", and searches were conducted using the Web of Science, PubMed, Google Scholar, and Scifinder databases. Publications from January 2020 to April 2025 were searched, excluding those focusing solely on extracts or review articles. In addition, previously reported studies on known PCSK9 inhibitors were excluded. The chemical structures were drawn by ChemDraw 23.1.2 software.

4. Conclusions and Future Perspectives

Since January 2020, concerted efforts have been made to isolate small molecules with PCSK9 inhibitory activity from natural products, based on research published on PCSK9 inhibitors during this period. As a result, a total of 350 compounds were reported between 2020 and April 2025, among which 57 compounds were confirmed to exhibit inhibitory activity against PCSK9. These active compounds were isolated from 18 plant species with diverse chemical profiles and include 19 isoprenoids, comprising 1 diterpene, 3 sesquiterpenes, and 15 triterpenes, along with 7 triterpenoidal saponins, 4 flavonoids, 9 flavonoid glycosides, 5 alkaloids, 1 isocoumarin, 5 lignans, 1 nucleic acid, 1 phenanthrene glycoside, 1 benzoquinone, 1 selaginellin derivative, and 3 stilbenes.

In addition, these isolated compounds modulate PCSK9 through various mechanisms, such as suppression of mRNA expression, downregulation of transcription regulators (HNF1α, SREBP2, FOXO1, and FOXO3), enhancement of LDLR activity, and inhibition of PCSK9-LDLR binding. Compared to earlier reviews, this study not only provides an updated overview of naturally derived compounds with proven PCSK9 inhibitory activity but also broadens the chemical diversity of PCSK9 inhibitors by introducing compound classes not previously reported. Overall, this review provides up-to-date information on naturally derived compounds with PCSK9 inhibitory activity and serves as a foundation for the development of drug candidates derived from natural products. However, most of the reported articles are limited to in vitro experiments using HepG2 cell lines, where the PCSK9 inhibitory effects were primarily assessed at the mRNA expression level. To further advance these compounds as viable therapeutic candidates, additional studies are needed, including in vivo investigations and/or the integration of RNA sequencing data with in silico approaches to identify potential interaction sites or direct target proteins [70,71]. These efforts are expected to provide deeper insights into the underlying mechanisms of PCSK9 inhibition.

Author Contributions: Conceptualization, Writing—original draft preparation and review, J.H., Writing—review, Supervision, H.K. All authors have read and agreed to the published version of the manuscript.

Funding: This work was supported by a National Research Foundation of Korea (NRF) grant funded by the Korean Government (MSIT), Grant RS-2023-00211868.

Institutional Review Board Statement: Not applicable.

Informed Consent Statement: Not applicable.

Data Availability Statement: No new data were created or analyzed in this study. Data sharing is not applicable to this article.

Conflicts of Interest: The authors declare no conflicts of interest.

Abbreviations

The following abbreviations are used in this manuscript: AUC, Area under the curve; PCSK9, Proprotein convertase subtilisin/Kexin type 9; SREBP, Sterol regulatory element-binding protein; LDL, Low-density lipoprotein; DiI-LDL, 1,1′-dioctadecyl-3,3,3′,3′-tetramethylindocarbocyanine perchlorate-labeled low-density lipoprotein; HNF-1α, Hepatocyte nuclear factor-1 alpha; FOXO1, Forkhead box protein O1; FOXO3, Forkhead box protein O3; EGCG, Epigallocatechin gallate; mRNA, Messenger RNA; IDOL, Inducible degrader of the low-density lipoprotein receptor; IC_{50}, Half-maximal inhibitory concentration; WST-8, Water-soluble tetrazolium 8; IP, Intraperitoneal; T/P ratio, Trough-to-peak ratio.

References

1. Huff, T.; Boyd, B.; Jialal, I. Physiology, Cholesterol. In *StatPearls*; StatPearls Publishing: Treasure Island, FL, USA, 2025.
2. High Cholesterol. Available online: https://www.hopkinsmedicine.org/health/conditions-and-diseases/high-cholesterol (accessed on 9 May 2025).
3. Dewick, P.M. *Medicinal Natural Products: A Biosynthetic Approach*, 3rd ed.; John Wiley & Sons: Hoboken, NJ, USA, 2009.
4. LDL and HDL Cholesterol and Triglycerides. Available online: https://www.cdc.gov/cholesterol/about/ldl-and-hdl-cholesterol-and-triglycerides.html (accessed on 9 May 2025).
5. Liu, Y.; Neogi, A.; Mani, A. The Role of Wnt Signalling in Development of Coronary Artery Disease and Its Risk Factors. *Open Biol.* **2020**, *10*, 200128. [CrossRef]
6. Linton, M.F.; Yancey, P.G.; Davies, S.S.; Jerome, W.G.; Linton, E.F.; Song, W.L.; Doran, A.C.; Vickers, K.C. The Role of Lipids and Lipoproteins in Atherosclerosis. In *Endotext [Internet]*; MDText.com, Inc.: South Dartmouth, MA, USA, 2019.
7. Stein, E.A.; Mellis, S.; Yancopoulos, G.D.; Stahl, N.; Logan, D.; Smith, W.B.; Lisbon, E.; Gutierrez, M.; Webb, C.; Wu, R.; et al. Effect of a Monoclonal Antibody to PCSK9 on LDL Cholesterol. *N. Engl. J. Med.* **2012**, *366*, 1108–1118. [CrossRef]
8. Cheng, J.M.; Oemrawsingh, R.M.; Garcia-Garcia, H.M.; Boersma, E.; van Geuns, R.-J.; Serruys, P.W.; Kardys, I.; Akkerhuis, K.M. PCSK9 in Relation to Coronary Plaque Inflammation: Results of the ATHEROREMO-IVUS Study. *Atherosclerosis* **2016**, *248*, 117–122. [CrossRef]
9. Goldstein, J.L.; Brown, M.S. The LDL Receptor. *Arterioscler. Thromb. Vasc. Biol.* **2009**, *29*, 431–438. [CrossRef]
10. Medina, M.W.; Krauss, R.M. The Role of HMGCR Alternative Splicing in Statin Efficacy. *Trends Cardiovasc. Med.* **2009**, *19*, 173–177. [CrossRef] [PubMed]
11. Maslub, M.G.; Radwan, M.A.; Daud, N.A.A.; Sha'aban, A. Association between CYP3A4/CYP3A5 Genetic Polymorphisms and Treatment Outcomes of Atorvastatin Worldwide: Is There Enough Research on the Egyptian Population? *Eur. J. Med. Res.* **2023**, *28*, 381. [CrossRef]
12. Maron, D.J.; Fazio, S.; Linton, M.F. Current Perspectives on Statins. *Circulation* **2000**, *101*, 207–213. [CrossRef] [PubMed]
13. Pang, J.; Chan, D.C.; Watts, G.F. The Knowns and Unknowns of Contemporary Statin Therapy for Familial Hypercholesterolemia. *Curr. Atheroscler. Rep.* **2020**, *22*, 64. [CrossRef]

14. Jang, A.Y.; Lim, S.; Jo, S.-H.; Han, S.H.; Koh, K.K. New Trends in Dyslipidemia Treatment. *Circ. J.* **2021**, *85*, 759–768. [CrossRef] [PubMed]
15. Seidah, N.G.; Benjannet, S.; Wickham, L.; Marcinkiewicz, J.; Jasmin, S.B.; Stifani, S.; Basak, A.; Prat, A.; Chretien, M. The Secretory Proprotein Convertase Neural Apoptosis-Regulated Convertase 1 (NARC-1): Liver Regeneration and Neuronal Differentiation. *Proc. Natl. Acad. Sci. USA* **2003**, *100*, 928–933. [CrossRef]
16. Bao, X.; Liang, Y.; Chang, H.; Cai, T.; Feng, B.; Gordon, K.; Zhu, Y.; Shi, H.; He, Y.; Xie, L. Targeting Proprotein Convertase Subtilisin/Kexin Type 9 (PCSK9): From Bench to Bedside. *Signal Transduct. Target. Ther.* **2024**, *9*, 13. [CrossRef]
17. Adorni, M.P.; Zimetti, F.; Lupo, M.G.; Ruscica, M.; Ferri, N. Naturally Occurring PCSK9 Inhibitors. *Nutrients* **2020**, *12*, 1440. [CrossRef]
18. Nassoury, N.; Blasiole, D.A.; Oler, A.T.; Benjannet, S.; Hamelin, J.; Poupon, V.; McPherson, P.S.; Attie, A.D.; Prat, A.; Seidah, N.G. The Cellular Trafficking of the Secretory Proprotein Convertase PCSK9 and Its Dependence on the LDLR. *Traffic* **2007**, *8*, 718–732. [CrossRef]
19. Qian, Y.-W.; Schmidt, R.J.; Zhang, Y.; Chu, S.; Lin, A.; Wang, H.; Wang, X.; Beyer, T.P.; Bensch, W.R.; Li, W.; et al. Secreted PCSK9 Downregulates Low Density Lipoprotein Receptor through Receptor-Mediated Endocytosis. *J. Lipid Res.* **2007**, *48*, 1488–1498. [CrossRef]
20. Denis, M.; Marcinkiewicz, J.; Zaid, A.; Gauthier, D.; Poirier, S.; Lazure, C.; Seidah, N.G.; Prat, A. Gene Inactivation of Proprotein Convertase Subtilisin/Kexin Type 9 Reduces Atherosclerosis in Mice. *Circulation* **2012**, *125*, 894–901. [CrossRef]
21. Tavori, H.; Melone, M.; Rashid, S. Alirocumab: PCSK9 Inhibitor for LDL Cholesterol Reduction. *Expert Rev. Cardiovasc. Ther.* **2014**, *12*, 1137–1144. [CrossRef]
22. Sabatine, M.S.; Giugliano, R.P.; Keech, A.C.; Honarpour, N.; Wiviott, S.D.; Murphy, S.A.; Kuder, J.F.; Wang, H.; Liu, T.; Wasserman, S.M.; et al. Evolocumab and Clinical Outcomes in Patients with Cardiovascular Disease. *N. Engl. J. Med.* **2017**, *376*, 1713–1722. [CrossRef] [PubMed]
23. Ridker, P.M.; Tardif, J.-C.; Amarenco, P.; Duggan, W.; Glynn, R.J.; Jukema, J.W.; Kastelein, J.J.P.; Kim, A.M.; Koenig, W.; Nissen, S.; et al. Lipid-Reduction Variability and Antidrug-Antibody Formation with Bococizumab. *N. Engl. J. Med.* **2017**, *376*, 1517–1526. [CrossRef]
24. Kastelein, J.J.P.; Nissen, S.E.; Rader, D.J.; Hovingh, G.K.; Wang, M.-D.; Shen, T.; Krueger, K.A. Safety and Efficacy of LY3015014, a Monoclonal Antibody to Proprotein Convertase Subtilisin/Kexin Type 9 (PCSK9): A Randomized, Placebo-Controlled Phase 2 Study. *Eur. Heart J.* **2016**, *37*, 1360–1369. [CrossRef] [PubMed]
25. Lin, J.; Ji, Y.; Wang, G.; Ma, X.; Yao, Z.; Han, X.; Chen, J.; Chen, J.; Huang, W.; Xu, G.; et al. Efficacy and Safety of Ongericimab in Chinese Patients with Heterozygous Familial Hypercholesterolemia: A Randomized, Double-Blind, Placebo-Controlled Phase 3 Trial. *Atherosclerosis* **2025**, *403*, 119120. [CrossRef] [PubMed]
26. Yan, S.; Zhao, X.; Xie, Q.; Du, W.; Ma, Q.; Zhu, T.; Deng, H.; Qian, L.; Zheng, S.; Cui, Y. Pharmacokinetic/LDL-C and Exposure-Response Analysis of Tafolecimab in Chinese Hypercholesterolemia Patients: Results from Phase I, II, and III Studies. *Clin. Transl. Sci.* **2023**, *16*, 2791–2803. [CrossRef]
27. Zhang, Y.; Pei, Z.; Chen, B.; Qu, Y.; Dong, X.; Yu, B.; Wang, G.; Xu, F.; Lu, D.; He, Z.; et al. Ebronucimab in Chinese Patients with Hypercholesterolemia—A Randomized Double-Blind Placebo-Controlled Phase 3 Trial to Evaluate the Efficacy and Safety of Ebronucimab. *Pharmacol. Res.* **2024**, *207*, 107340. [CrossRef]
28. Sun, Y.; Lv, Q.; Guo, Y.; Wang, Z.; Huang, R.; Gao, X.; Han, Y.; Yao, Z.; Zheng, M.; Luo, S.; et al. Recaticimab as Add-on Therapy to Statins for Nonfamilial Hypercholesterolemia: The Randomized, Phase 3 REMAIN-2 Trial. *J. Am. Coll. Cardiol.* **2024**, *84*, 2037–2047. [CrossRef]
29. Turner, T.A.; Butcher, B.; Mangu, P.; Kereiakes, D.; Fu, R.; Bakker-Arkema, R.; Stein, E.A. Results of a 52 Week Open-Label Phase 2B Study to Assess Long-Term Safety, Immunogenicity and LDL-C Efficacy of Monthly Dosing with LIB003 a Novel Anti-PCSK9 Recombinant Fusion Protein. *Atherosclerosis* **2020**, *315*, e9. [CrossRef]
30. Burnett, J.R.; Hooper, A.J. MK-0616: An Oral PCSK9 Inhibitor for Hypercholesterolemia Treatment. *Expert Opin. Investig. Drugs* **2023**, *32*, 873–878. [CrossRef]
31. Koren, M.J.; Descamps, O.; Hata, Y.; Hengeveld, E.M.; Hovingh, G.K.; Ikonomidis, I.; Radu Juul Jensen, M.D.; Langbakke, I.H.; Martens, F.M.A.C.; Søndergaard, A.L.; et al. PCSK9 Inhibition with Orally Administered NNC0385-0434 in Hypercholesterolaemia: A Randomised, Double-Blind, Placebo-Controlled and Active-Controlled Phase 2 Trial. *Lancet Diabetes Endocrinol.* **2024**, *12*, 174–183. [CrossRef]
32. Raal, F.J.; Mehta, V.; Kayikcioglu, M.; Blom, D.; Gupta, P.; Elis, A.; Turner, T.; Daniels, C.; Vest, J.; Mitchell, T.; et al. Lerodalcibep and Evolocumab for the Treatment of Homozygous Familial Hypercholesterolaemia with PCSK9 Inhibition (LIBerate-HoFH): A Phase 3, Randomised, Open-Label, Crossover, Non-Inferiority Trial. *Lancet Diabetes Endocrinol.* **2025**, *13*, 178–187. [CrossRef] [PubMed]

33. Mansoor, T.; Rao, B.H.; Gupta, K.; Parikh, S.S.; Abramov, D.; Mehta, A.; Al Rifai, M.; Virani, S.S.; Nambi, V.; Minhas, A.M.K.; et al. Inclisiran as a SiRNA Inhibitor of Proprotein Convertase Subtilisin/Kexin Type 9 (PCSK9); Past, Present, and Future. *Am. J. Cardiovasc. Drugs* **2025**, *25*, 293–306. [CrossRef] [PubMed]
34. Vega, R.; Garkaviy, P.; Knöchel, J.; Barbour, A.; Rudvik, A.; Laru, J.; Twaddle, L.; Mccarthy, M.C.; Rosenmeier, J.B. AZD0780, the First Oral Small Molecule PCSK9 Inhibitor for the Treatment of Hypercholesterolemia: Results from a Randomized, Single-Blind, Placebo-Controlled Phase 1 Trial. *Atherosclerosis* **2024**, *395*, 118514. [CrossRef]
35. Mitchell, T.; Chao, G.; Sitkoff, D.; Lo, F.; Monshizadegan, H.; Meyers, D.; Low, S.; Russo, K.; DiBella, R.; Denhez, F.; et al. Pharmacologic Profile of the Adnectin BMS-962476, a Small Protein Biologic Alternative to PCSK9 Antibodies for Low-Density Lipoprotein Lowering. *J. Pharmacol. Exp. Ther.* **2014**, *350*, 412–424. [CrossRef]
36. Aguchem, R.N.; Okagu, I.U.; Okorigwe, E.M.; Uzoechina, J.O.; Nnemolisa, S.C.; Ezeorba, T.P.C. Role of CETP, PCSK-9, and CYP7-Alpha in Cholesterol Metabolism: Potential Targets for Natural Products in Managing Hypercholesterolemia. *Life Sci.* **2024**, *351*, 122823. [CrossRef]
37. Barale, C.; Melchionda, E.; Morotti, A.; Russo, I. PCSK9 Biology and Its Role in Atherothrombosis. *Int. J. Mol. Sci.* **2021**, *22*, 5880. [CrossRef] [PubMed]
38. Hummelgaard, S.; Vilstrup, J.P.; Gustafsen, C.; Glerup, S.; Weyer, K. Targeting PCSK9 to Tackle Cardiovascular Disease. *Pharmacol. Ther.* **2023**, *249*, 108480. [CrossRef]
39. Waiz, M.; Alvi, S.S.; Khan, M.S. Potential Dual Inhibitors of PCSK-9 and HMG-R from Natural Sources in Cardiovascular Risk Management. *EXCLI J.* **2022**, *21*, 47–76. [PubMed]
40. Liu, C.; Chen, J.; Chen, H.; Zhang, T.; He, D.; Luo, Q.; Chi, J.; Hong, Z.; Liao, Y.-J.; Zhang, S.; et al. PCSK9 Inhibition: From Current Advances to Evolving Future. *Cells* **2022**, *11*, 2972. [CrossRef]
41. Singh, S.; Sharma, H.; Ramankutty, R.; Ramaswamy, S.; Agrawal, N. Natural Proprotein Convertase Subtilisin/Kexin Type 9 Inhibitors: A Review. *Comb. Chem. High Throughput Screen.* **2023**, *26*, 2668–2678. [CrossRef]
42. Liou, J.-W.; Chen, P.-Y.; Gao, W.-Y.; Yen, J.-H. Natural Phytochemicals as Small-Molecule Proprotein Convertase Subtilisin/Kexin Type 9 Inhibitors. *Tzu Chi Med. J.* **2024**, *36*, 360–369. [CrossRef]
43. Li, H.-H.; Li, J.; Zhang, X.-J.; Li, J.-M.; Xi, C.; Wang, W.-Q.; Lu, Y.-L.; Xuan, L.-J. 23,24-Dihydrocucurbitacin B Promotes Lipid Clearance by Dual Transcriptional Regulation of LDLR and PCSK9. *Acta Pharmacol. Sin.* **2020**, *41*, 327–335. [CrossRef]
44. Zhang, X.; Li, H.; Wang, W.; Chen, T.; Xuan, L. Lipid-Lowering Activities of Cucurbitacins Isolated from Trichosanthes Cucumeroides and Their Synthetic Derivatives. *J. Nat. Prod.* **2020**, *83*, 3536–3544. [CrossRef]
45. Pel, P.; Chae, H.-S.; Nhoek, P.; Kim, Y.-M.; Khiev, P.; Kim, G.J.; Nam, J.-W.; Choi, H.; Choi, Y.H.; Chin, Y.-W. A Stilbene Dimer and Flavonoids from the Aerial Parts of Chromolaena Odorata with Proprotein Convertase Subtilisin/Kexin Type 9 Expression Inhibitory Activity. *Bioorg. Chem.* **2020**, *99*, 103869. [CrossRef] [PubMed]
46. Li, H.; Dong, B.; Park, S.W.; Lee, H.-S.; Chen, W.; Liu, J. Hepatocyte Nuclear Factor 1alpha Plays a Critical Role in PCSK9 Gene Transcription and Regulation by the Natural Hypocholesterolemic Compound Berberine. *J. Biol. Chem.* **2009**, *284*, 28885–28895. [CrossRef] [PubMed]
47. Nhoek, P.; Chae, H.-S.; Kim, Y.-M.; Pel, P.; Huh, J.; Kim, H.W.; Choi, Y.H.; Lee, K.; Chin, Y.-W. Sesquiterpenoids from the Aerial Parts of Salvia Plebeia with Inhibitory Activities on Proprotein Convertase Subtilisin/Kexin Type 9 Expression. *J. Nat. Prod.* **2021**, *84*, 220–229. [CrossRef] [PubMed]
48. Weng, X.; Lou, Y.-Y.; Wang, Y.-S.; Huang, Y.-P.; Zhang, J.; Yin, Z.-Q.; Pan, K. New Dammarane-Type Glycosides from Gynostemma Pentaphyllum and Their Lipid-Lowering Activity. *Bioorg. Chem.* **2021**, *111*, 104843. [CrossRef]
49. Kim, E.; Kim, Y.-M.; Ahn, J.; Chae, H.-S.; Chin, Y.-W.; Kim, J. Prenylated Flavonoid Glycosides with PCSK9 MRNA Expression Inhibitory Activity from the Aerial Parts of Epimedium Koreanum. *Molecules* **2021**, *26*, 3590. [CrossRef]
50. Woo, S.; Chae, H.-S.; Kim, J.; Chin, Y.-W. Selaginellin Derivatives from Selaginella Tamariscina and Their Upregulating Effects on Low-Density Lipoprotein Receptor Expression. *J. Nat. Prod.* **2021**, *84*, 857–864. [CrossRef]
51. Ahn, J.; Chae, H.-S.; Pel, P.; Kim, Y.-M.; Choi, Y.H.; Kim, J.; Chin, Y.-W. Dilignans with a Chromanol Motif Discovered by Molecular Networking from the Stem Barks of Magnolia Obovata and Their Proprotein Convertase Subtilisin/Kexin Type 9 Expression Inhibitory Activity. *Biomolecules* **2021**, *11*, 463. [CrossRef] [PubMed]
52. Huang, Y.-P.; Wang, Y.-S.; Liu, B.-W.; Song, Z.; Liang, X.-S.; Teng, Y.; Zhang, J.; Yin, Z.-Q.; Pan, K. Dammarane-Type Saponins with Proprotein Convertase Subtilisin/Kexin Type 9 Inhibitory Activity from Gynostemma Pentaphyllum. *Phytochemistry* **2022**, *194*, 113005. [CrossRef]
53. Zhang, J.; Yan, X.; Jin, Q.; Chen, J.; Yang, L.; Wei, W.; Qu, H.; Yao, C.; Hou, J.; Gong, L.; et al. Novel Triterpenoids from Alisma Plantago-Aquatica with Influence on LDL Uptake in HepG2 Cells by Inhibiting PCSK9. *Phytomedicine* **2022**, *105*, 154342. [CrossRef]

54. Huh, J.; Park, T.K.; Chae, H.-S.; Nhoek, P.; Kim, Y.-M.; An, C.-Y.; Lee, S.; Kim, J.; Chin, Y.-W. Acylated Saponins and Flavonoid Glycosides from the Fruits of Stewartia Koreana. *Phytochemistry* **2022**, *193*, 112980. [CrossRef]
55. Pel, P.; Kim, Y.-M.; Kim, H.J.; Nhoek, P.; An, C.-Y.; Son, M.-G.; Won, H.; Lee, S.E.; Lee, J.; Kim, H.W.; et al. Isocoumarins and Benzoquinones with Their Proprotein Convertase Subtilisin/Kexin Type 9 Expression Inhibitory Activities from Dried Roots of Lysimachia Vulgaris. *ACS Omega* **2022**, *7*, 47296–47305. [CrossRef]
56. Pel, P.; Chae, H.-S.; Nhoek, P.; Kim, Y.-M.; An, C.-Y.; Yoo, H.; Kang, M.; Kim, H.W.; Choi, Y.H.; Chin, Y.-W. Chemical Constituents from the Roots and Rhizomes of Sophora Tonkinensis and Their Effects on Proprotein Convertase Substilisin/Kexin Type 9 Expression. *ACS Omega* **2022**, *7*, 20952–20958. [CrossRef]
57. Nhoek, P.; An, C.-Y.; Son, M.-G.; Chae, H.-S.; Pel, P.; Kim, Y.-M.; Khiev, P.; Choi, W.J.; Choi, Y.H.; Chin, Y.-W. Stereochemical Assignment of Clerodane-Type Diterpenes from the Fruits of Casearia Grewiifolia and Their Ability to Inhibit PCSK9 Expression. *Phytochemistry* **2023**, *216*, 113864. [CrossRef]
58. Wang, J.-Q.; Lin, Z.-C.; Li, L.-L.; Zhang, S.-F.; Li, W.-H.; Liu, W.; Song, B.-L.; Luo, J. SUMOylation of the Ubiquitin Ligase IDOL Decreases LDL Receptor Levels and Is Reversed by SENP1. *J. Biol. Chem.* **2021**, *296*, 100032. [CrossRef]
59. Martinelli, L.; Adamopoulos, A.; Johansson, P.; Wan, P.T.; Gunnarsson, J.; Guo, H.; Boyd, H.; Zelcer, N.; Sixma, T.K. Structural Analysis of the LDL Receptor-Interacting FERM Domain in the E3 Ubiquitin Ligase IDOL Reveals an Obscured Substrate-Binding Site. *J. Biol. Chem.* **2020**, *295*, 13570–13583. [CrossRef]
60. An, C.-Y.; Son, M.-G.; Chin, Y.-W. Acyclic Triterpenoids from Alpinia Katsumadai Seeds with Proprotein Convertase Subtilisin/Kexin Type 9 Expression and Secretion Inhibitory Activity. *ACS Omega* **2023**, *8*, 32804–32816. [CrossRef]
61. Hu, X.; Chen, T.; Guo, P.; Wang, Q.; Ding, A.; Qin, G.; Wang, W.; Xuan, L. Amide Alkaloids and Neolignans from Piper Hongkongense and Their Inhibitory Activities of PCSK9 Expression. *Fitoterapia* **2024**, *175*, 105951. [CrossRef]
62. Son, M.-G.; Pel, P.; An, C.-Y.; Park, C.-W.; Lee, S.H.; Yang, T.-J.; Chin, Y.-W. Chemical Constituents from the Roots of Cynanchum Wilfordii with PCSK9 Secretion Inhibitory Activities. *Phytochemistry* **2024**, *226*, 114205. [CrossRef] [PubMed]
63. Won, H.; Son, M.-G.; Pel, P.; Nhoek, P.; An, C.-Y.; Kim, Y.-M.; Chae, H.-S.; Chin, Y.-W. Chemical Constituents from Morus Alba with Proprotein Convertase Subtilisin/Kexin Type 9 Expression and Secretion Inhibitory Activity. *Org. Biomol. Chem.* **2023**, *21*, 2801–2808. [CrossRef] [PubMed]
64. Dong, B.; Wu, M.; Li, H.; Kraemer, F.B.; Adeli, K.; Seidah, N.G.; Park, S.W.; Liu, J. Strong Induction of PCSK9 Gene Expression through HNF1alpha and SREBP2: Mechanism for the Resistance to LDL-Cholesterol Lowering Effect of Statins in Dyslipidemic Hamsters. *J. Lipid Res.* **2010**, *51*, 1486–1495. [CrossRef] [PubMed]
65. Ataei, S.; Kesharwani, P.; Sahebkar, A. Berberine: Ins and Outs of a Nature-Made PCSK9 Inhibitor. *EXCLI J.* **2022**, *21*, 1099–1110.
66. Wang, X.; Chen, X.; Zhang, X.; Su, C.; Yang, M.; He, W.; Du, Y.; Si, S.; Wang, L.; Hong, B. A Small-Molecule Inhibitor of PCSK9 Transcription Ameliorates Atherosclerosis through the Modulation of FoxO1/3 and HNF1α. *EBioMedicine* **2020**, *52*, 102650. [CrossRef]
67. Wei, Y.; Wen, H.; Yang, L.; Zhang, B.; Li, X.; Li, S.; Dong, J.; Liang, Z.; Zhang, Y. Hypecotumines A-D, New Isoquinoline Alkaloids with Potential PCSK9 Inhibition Activity from *Hypecoum erectum* L. *Nat. Prod. Bioprospect.* **2024**, *14*, 57.
68. Lee, S.; Son, M.-G.; Kim, Y.-M.; An, C.-Y.; Kim, H.J.; Nhoek, P.; Pel, P.; Won, H.; Lee, Y.; Yun, N.; et al. Dihydrostilbenes and Flavonoids from Whole Plants of Jacobaea Vulgaris. *Phytochemistry* **2024**, *222*, 114107. [CrossRef] [PubMed]
69. An, C.-Y.; Pel, P.; Bae, M.; Park, C.-W.; Kwon, H.; Lee, H.S.; Van Dung, L.; Kim, C.; Lee, D.; Choi, Y.H.; et al. Cycloartane-Type Triterpenoids from Combretum Quadrangulare Kurz with PCSK9 Secretion Inhibitory Activities. *Phytochemistry* **2025**, *230*, 114330. [CrossRef] [PubMed]
70. Siddiq, A.A.; Dileep, S.A.; Sj, A.R.; Singam, S.S.R.; Martin, A. Saffron and Its Active Constituents Ameliorate Hypercholesterolemia by Inhibiting PCSK9 and Modulating Sortilin, LDLR, and SREBP-2 Signaling in High Fat Diet Induced Hypercholesterolemic C57BL/6 Mice. *J. Ethnopharmacol.* **2025**, *346*, 119697. [CrossRef] [PubMed]
71. Rho, H.; Terry, A.R.; Chronis, C.; Hay, N. Hexokinase 2-Mediated Gene Expression via Histone Lactylation Is Required for Hepatic Stellate Cell Activation and Liver Fibrosis. *Cell Metab.* **2023**, *35*, 1406–1423.e8. [CrossRef]

molecules MDPI

Review

Recent Advances in the Application of Nitro(het)aromatic Compounds for Treating and/or Fluorescent Imaging of Tumor Hypoxia

Kameliya Anichina [1], Nikolay Lumov [1,2], Ventsislav Bakov [1], Denitsa Yancheva [1,2] and Nikolai Georgiev [1,*]

1 Department of Organic Synthesis, University of Chemical Technology and Metallurgy, 8 Kliment Ohridski Blvd., 1756 Sofia, Bulgaria; kameliya_anichina@uctm.edu (K.A.); nikolay.lumov@orgchm.bas.bg (N.L.); bakov@uctm.edu (V.B.); denitsa.pantaleeva@orgchm.bas.bg (D.Y.)

2 Institute of Organic Chemistry with Centre of Phytochemistry, Bulgarian Academy of Sciences, Acad. G. Bonchev str. Bl. 9, 1113 Sofia, Bulgaria

* Correspondence: nikigeorgiev@uctm.edu

Abstract: This review delves into recent advancements in the field of nitro(het)aromatic bioreductive agents tailored for hypoxic environments. These compounds are designed to exploit the low-oxygen conditions typically found in solid tumors, making them promising candidates for targeted cancer therapies. Initially, this review focused on their role as gene-directed enzyme prodrugs, which are inert until activated by specific enzymes within tumor cells. Upon activation, these prodrugs undergo chemical transformations that convert them into potent cytotoxic agents, selectively targeting cancerous tissue while sparing healthy cells. Additionally, this review discusses recent developments in prodrug conjugates containing nitro(het)aromatic moieties, designed to activate under low-oxygen conditions within tumors. This approach enhances their efficacy and specificity in cancer treatment. Furthermore, this review covers innovative research on using nitro(het)aromatic compounds as fluorescent probes for imaging hypoxic tumors. These probes enable non-invasive visualization of low-oxygen regions within tumors, providing valuable insights for the diagnosis, treatment planning, and monitoring of therapeutic responses. We hope this review will inspire researchers to design and synthesize improved compounds for selective cancer treatment and early diagnostics.

Keywords: nitroaromatic compounds; hypoxia; tumor treatment; fluorescence imaging

1. Introduction

Hypoxia, a condition of low oxygen levels ranging from 0.02 to 2%, is a physiological characteristic of most solid tumors. It occurs because the tumor's rapid growth outstrips the oxygen supply and is compounded by impaired blood flow due to the formation of abnormal blood vessels supplying the tumor [1].

Severe hypoxia (<0.5% O_2) suppresses energy-consuming processes in the cells such as translation and disulfide bond formation, causing protein misfolding and activating the unfolded protein response (UPR). The UPR inhibits global protein synthesis while selectively translating mRNAs to maintain endoplasmic reticulum homeostasis and promote hypoxia tolerance. Furthermore, under severe hypoxia, the ataxia–telangiectasia mutated (ATM) gene activates and DNA repair pathways are downregulated, decreasing RAD_{51} expression and impairing homologous recombination. Reoxygenation after extreme hypoxia causes DNA damage and genomic instability, increasing mutation rates and metastatic potential. Severe hypoxia also hinders the repair of G1-associated DNA double-strand breaks in irradiated cells, increasing genomic instability [2].

Cell survival strategies under severe hypoxia include suppressing apoptosis, initiating angiogenesis (the formation of new but often abnormal blood vessels) and erythropoiesis (the formation of red blood cells), and shifting cell metabolism from oxidative phosphorylation to glycolysis [3]. This leads to increased tumor proliferation and local invasiveness. Additionally, in the absence of oxygen, tumor cells influence immune responses and initiate rapid DNA damage repair mechanisms to maintain growth and survival, ultimately becoming resistant to radiation and chemotherapy [4,5].

The unique microenvironment of hypoxic tumors represents an opportunity for targeted therapy through the development of bioreductive drugs (BDs), named later hypoxia-activated prodrugs (HAPs). They are prodrugs that undergo biotransformation to cytotoxic compounds under conditions of low oxygen tension and in the presence of high levels of specific reductases [6,7]. This strategy, which originated in the early 1970s [8], allows the selective killing of cancer cells while exhibiting minimal or no toxicity to normal cells and well-oxygenated cancer cells.

Currently, four distinct chemical entities are known to selectively target hypoxic cells: nitro(hetero)cyclic compounds, aromatic N-oxides, aliphatic N-oxides, quinones, and transition metal complexes [9]. Among them, nitroaryl- and nitroheteroaryl-based compounds show immense potential. These compounds act as bioreductive components within the chemical architecture of bioactive molecules and as triggers capable of inducing the release of the chemotherapeutic agent within hypoxic tumor regions [10].

The structures of several bioreductive agents containing nitro groups, which have been evaluated in clinical trials, are presented in Figure 1.

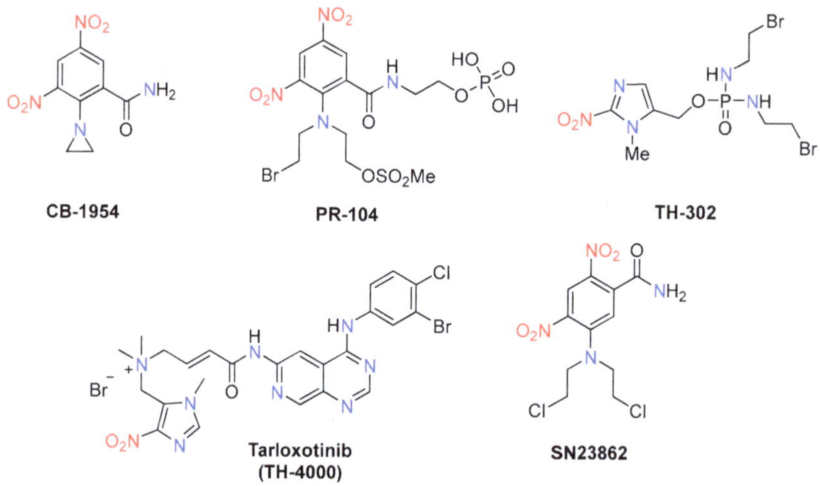

Figure 1. Chemical structures of some hypoxia-activated nitro(het)aromatic prodrugs.

Despite the promising results from clinical trials of the HAPs, none have yet been commercialized. The failure in clinical trials is partly due to the following reasons: insufficient selectivity and efficacy for hypoxic tumor cells over normoxic healthy cells and potential toxicity; the highly heterogeneous hypoxic tumor environments; pharmacokinetics and stability issues etc. [11]. For example, TH-302 (evofosfamide, (1-methyl-2-nitro-1*H*-imidazol-5-yl)methyl *N*,*N*′-bis(2-bromoethyl)phosphorodiamidate), a widely studied HAP, showed limited clinical benefits in a phase I trial for advanced solid tumors, due to its insufficient selectivity for hypoxic regions and associated skin and mucosal toxicity [12]. Skin and mucosal toxicity, along with bone marrow suppression, were the most common toxicities observed when evaluating the therapeutic potential of TH-302 combined with gemcitabine for pancreatic cancer [13]. In the phase III multicenter clinical trial (TH CR-406/SARC021), 640 patients with soft tissue sarcoma were enrolled to assess the efficacy of combining TH-302 (300 mg/m^2) with doxorubicin (75 mg/m^2). The results showed that this combination did not improve overall survival compared to doxorubicin alone [14].

Similarly, the prodrug PR-104 (2-({2-[(2-bromoethyl)[2-(methanesulfonyloxy)-ethyl]amino]-3,5-dinitrophenyl}formami-do)etho-xy]phosphonic acid), another HAP, showed inadequate efficacy and raised safety concerns in phase II trials for relapsed or refractory acute myeloid leukemia. The most frequent treatment-related grade 3/4 adverse events were myelosuppression (anemia 62%, neutropenia 50%, thrombocytopenia 46%), febrile neutropenia (40%), infections (24%), and enterocolitis (14%) [15]. The combination of PR-104 with chemotherapeutics such as gemcitabine or docetaxel in advanced solid tumors was discontinued due to dose-limiting thrombocytopenia [16]. This necessitates the development of new nitroaryl- and/or nitroheteroaryl-based HAPs with enhanced specificity for hypoxic cells, minimized off-target effects. Additionally, improvements in the pharmacokinetic properties of HAPs via chemical modifications or formulation strategies are required.

In this review, we explore recent advancements in the design and synthesis of nitro(het)aromatic bioreductive agents, specifically focusing on their developments from 2018 to 2024. We first highlight their potential as gene-directed enzyme prodrugs that undergo controlled conversion into cytotoxic agents within target cells. Additionally, we examine the progress made in developing bioreductive-activated prodrug conjugates (BAPCs), which contain nitro(het)aromatic moieties (triggers) that activate drug molecules in the low-oxygen environment of tumors. The ability of nitro(het)aromatic compounds to undergo bioreductive activation under hypoxic conditions makes them highly effective for non-invasive tumor visualization and monitoring. We also discuss recent research on nitroaromatic compounds as tools for the fluorescent imaging of hypoxic tumors. We hope this review will aid researchers in designing new nitro(het)aromatic structures for treating and/or fluorescently imaging hypoxic tumors, paving the way for improved cancer diagnostics and treatment monitoring.

2. Nitro(het)aromatic Bioreductive Agents for Use in Gene-Directed Enzyme Prodrug Therapy

The mechanism by which nitro compounds localize within cells under hypoxic conditions involves multiple stages (Figure 2). In the initial and most crucial step, cellular nitroreductase enzymes reduce the nitro group of the prodrug molecule to a nitro anion radical, a short-lived species, especially in an aqueous medium. Under normal oxygen levels (normoxia), the $NO_2^{\cdot-}$ radical is quickly oxidized back to the original NO_2 group (re-oxygenated), producing superoxide anions ($O_2^{\cdot-}$) in the process [17].

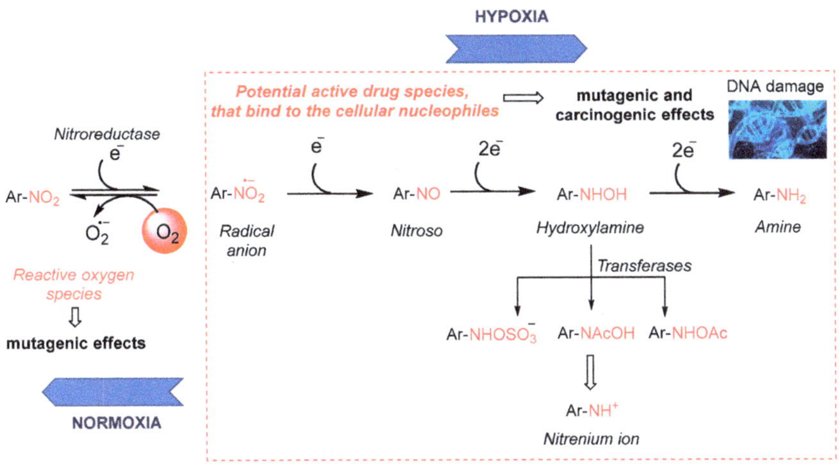

Figure 2. Bioreduction of nitroaryl or heteroaryl compounds under hypoxic conditions.

In the absence of oxygen, the prodrug radical anion undergoes stepwise reduction to nitroso, hydroxylamine, and amine (Figure 2). These reactive intermediates bind to cellular macromolecules such as DNA, proteins, and glutathione, thereby persisting in oxygen-deprived tissues and exhibiting mutagenic and carcinogenic effects [18]. The toxicity of nitro compounds, both desired and undesired, is linked to each of these intermediates. Hydroxylamine derivatives can cause methemoglobinemia, while nitro radical anions, nitroso derivatives, and esterified hydroxylamines (e.g., sulfo derivatives) promote mutagenic and carcinogenic effects. Additionally, superoxide anions, hydrogen peroxide, and hydroxyl radicals generated during the reduction process can also have mutagenic effects [18]. For example, the water-soluble phosphate prodrug PR-104 (Figure 1) is rapidly hydrolyzed by phosphatases in vivo to the less soluble alcohol metabolite. The latter is sufficiently lipophilic to penetrate through multiple layers of tumor cells and to reach the hypoxic target cells. There, a nitro group in the para position to the mustard residue of the prodrug undergoes reduction to hydroxylamine and an amine group, respectively. These two cytotoxic metabolites act as DNA interstrand cross-linking agents, able to diffuse locally and kill neighboring cells [19]. The bioreductive metabolic pathways of the other nitro(het)aromatic prodrugs, shown in Figure 1, are detailed in [8].

One of the most significant applications of nitro(het)aromatic compounds, in combination with enzymes from the nitroreductase group, is gene-directed enzyme prodrug therapy (GDEPT), widely explored in chemotherapy [20–23]. This therapy is a variation of direct enzyme prodrug therapy (DEPT). In DEPT, an exogenous converting enzyme is delivered to the tumor cell, rendering the cell sensitive to the administered prodrug. The advantage of using exogenous enzymes lies in their ability to activate substances that are inert to human enzymes, thereby minimizing off-target effects. However, the challenge lies in delivering these enzymes specifically to tumor cells.

There are two primary approaches to enzyme prodrug therapy. The first approach involves directly delivering the enzyme linked to a tumor-targeting molecule, ensuring the enzyme reaches the cancer cells. The second approach, GDEPT, involves a more indirect strategy. GDEPT utilizes gene therapy techniques to introduce genes encoding the prodrug-activating enzyme specifically into the tumor cells [20,23]. Once inside the tumor cells, these genes are expressed, producing the enzyme that can then convert the administered prodrug into its active, cytotoxic form within the tumor microenvironment. Different nitroreductases derived from bacteria such as *Escherichia coli*, *Pseudomonas pseudoalcaligenes*, and *Staphylococcus saprophyticus* have been utilized in these studies. The GDEPT approach works through three key steps: 1. Genes encoding the nitroreductase enzyme are delivered to the tumor cells (Figure 3A). This targeting can be achieved using vectors such as viruses, plasmids, or nanoparticles engineered to selectively infect or enter tumor cells. 2. Once inside the tumor cells, the genes are transcribed and translated to produce the nitroreductase enzyme (Figure 3B). 3. The administered nitro(het)aromatic prodrug is then activated by this enzyme within the tumor cells, converting it into a cytotoxic agent that induces cell death (Figure 3C).

The specificity of GDEPT allows for high concentrations of the cytotoxic agent to be generated directly within the tumor, minimizing the damage to surrounding healthy tissues. Over the past twenty years, the GDEPT approach has seen significant progress, with numerous enzyme/prodrug systems proving effective in preclinical and clinical studies. Nonetheless, considerable efforts are still required to fully harness the potential of this promising cancer treatment option [24].

Prof. Ay. M.'s research group has worked on discovering new and effective nitroreductase—prodrug combinations for use in cancer therapy. For this purpose, they characterized a new nitroreductase, Ssap-NtrB (*Staphylococcus saprophyticus* supsp. saprophyticus), in 2012 [25], synthesized various nitro functional group-containing prodrug candidates, and investigated their enzymatic and cytotoxic effects on different cancer cells.

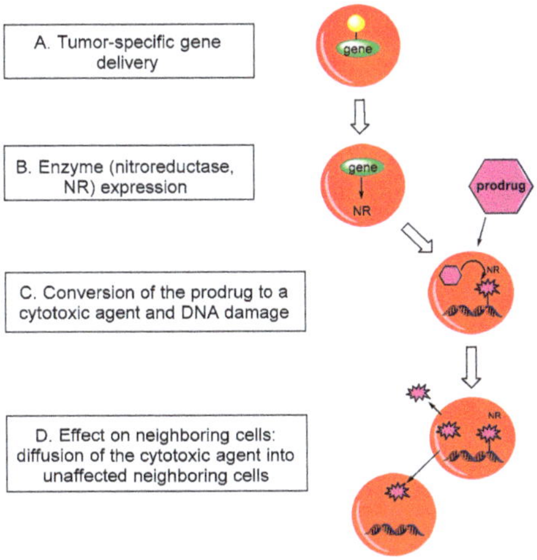

Figure 3. Principle of operation of GDEPT.

Güngör T. et al. designed prodrugs **1–4** based on the model prodrug CB1954 and some benzamides. Their concept involved replacing the -$CONH_2$ group of CB1954 with the nitro group containing -CONHAr groups (Figure 4) [26]. According to the HPLC results, all prodrugs were activated by Ssap-NtrB. Prodrugs **1** and **2** produced one metabolite each, while prodrugs **3** and **4** produced three metabolites each upon activation. Derivative **3** was significantly more active than CB1954 and SN23862, with 137- and 31-fold higher activity for Ssap-NtrB, respectively. Among all the prodrug metabolites following Ssap-NtrB reduction, *N*-(2,4-dinitrophenyl)-4-nitrobenzamide **3** was notably effective and toxic to PC3 cells, comparable to CB1954. Kinetic parameters, molecular docking, and the HPLC results also indicated that prodrug 3 interacts more favorably with Ssap-NtrB than prodrugs 1, 2, and 4, or the known cancer prodrugs CB1954 and SN23862. This makes prodrug 3 a promising candidate for NTR-based cancer therapy.

The metabolites of prodrugs **6a** and **6b** exhibited IC_{50} values of 1.806 nM and 1.808 nM, respectively [27]. The metabolite of prodrug **8a** demonstrated an IC_{50} value of 1.793 nM, comparable to CB1954. The common structural feature of the most active nitrobenzamide compounds (**6a**, **6b**, and **8a**) includes a nitro group in p-position on the phenyl core relative to the amide group and the presence of nitrogen-containing heterocyclic systems such as piperidine (**6a**), morpholine (**6b**), or a saturated 1,4-cyclohexyl moiety (**8a**). The compounds with two phenyl nuclei and one amide group exhibited the highest toxicity, followed by bis-benzamides. The toxicity of the tested benzamides can be ranked in the following order: **6** < **5** < **8** < **7**. As a result of theoretical and biological studies, combinations of **6a**, **6b**, and **8a** with Ssap-NtrB can be suggested as potential prodrugs–enzyme combinations at nitroreductase-based cancer therapy, compared with the CB1954–NfsB combination.

Further in this direction, Tokay et al. present the synthesis of *N*-(substituted)-2,4-dinitroaniline derivatives, in particular symmetrical bis(2,4-dinitrophenyl)diamine derivatives **9** and *N*-(5-morpholino-2,4-dinitro phenyl)alkanamides **10**. These aromatic secondary amines were derived from 2,4-dinitro-1-chlorobenzene and various aliphatic, alicyclic, aromatic, or heterocyclic diamino derivatives utilizing Et_3N or NaH as a base in DMF solvent at room temperature or 60–70 °C via the SNAr reaction mechanism [28]. The design of these compounds was based on model bioreductive dinitroaniline prodrugs such as CB1954, SN23862, and PR-104A. The cytotoxic effects of prodrug candidates were assessed using the MTT assay on human hepatoma cells (Hep3B), prostate cancer cells (PC-3), and human umbilical vein endothelial cells (HUVEC) as healthy controls. The compounds with minimal toxicity were further investigated to evaluate their potential as prodrug candidates. Biochemical analyses were conducted to examine the reduction profiles and kinetics of prodrug–Ssap-NtrB combinations. Subsequently, selected prodrug–Ssap-NtrB combinations were applied to prostate cancer cells to assess their toxicity. The combined results from theoretical, in vitro cytotoxic, and biochemical studies indicate that prodrug/enzyme combinations such as **9a**—Ssap-NtrB, **9b**—Ssap-NtrB, and **10**—Ssap-NtrB hold promise as potential candidates for nitroreductase (Ntr)-based prostate cancer therapy.

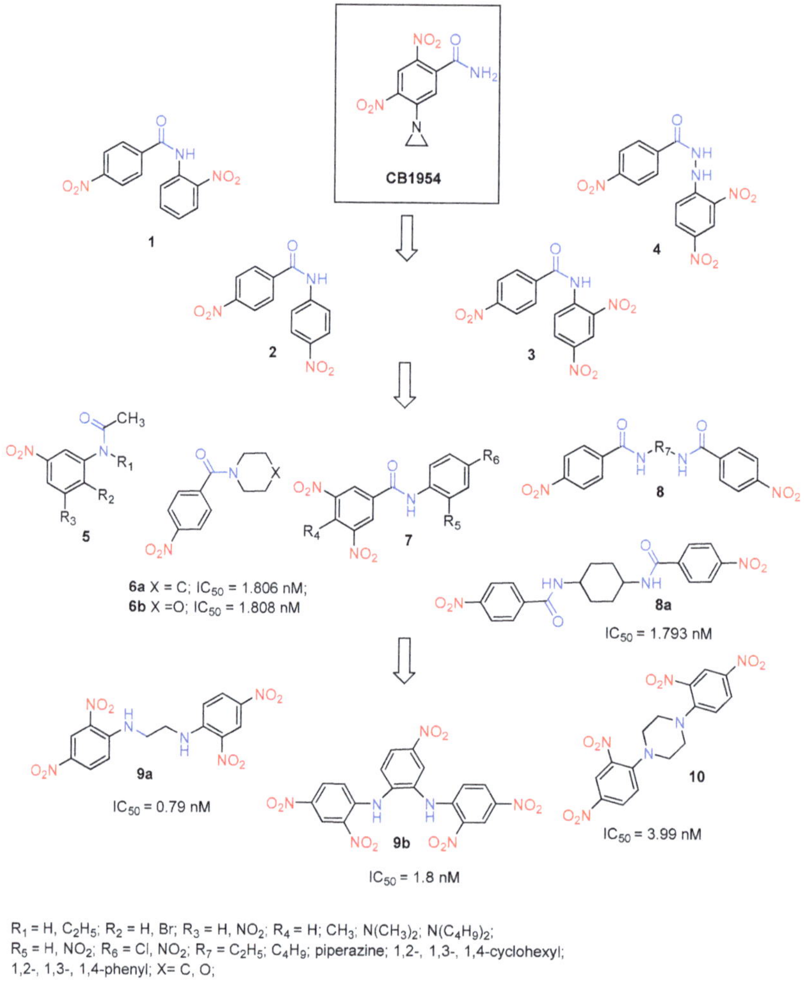

Figure 4. Structures of prodrug candidates **1–10** and a half-maximal inhibitory concentration (IC_{50}) of their metabolites.

Building on their previous works, Güngör and colleagues synthesized a series of *N*-heterocyclic nitro prodrugs (**11–13**, Figure 5) containing pyrimidine, triazine, and piperazine rings. Nitro-containing triazine derivatives **11** and **12** are synthesized from cyanuric chloride and the amines via a nucleophilic substitution reaction. The process involves refluxing cyanuric chloride and aromatic amines in acetic acid for varying reaction times (15 min to 24 h), followed by purification through crystallization with isopropyl alcohol. For the synthesis of urea derivatives of nitrophenyls and piperazine **13a–b**, a Curtius rearrangement is applied. This involves first reacting nitrobenzoyl chlorides with sodium azide to obtain nitrobenzoyl azide derivatives. Subsequently, at the reflux temperature of toluene, nitrophenyl isocyanate forms as an unstable intermediate, which then reacts with piperazine, yielding the desired products with high efficiency (78–96%). Prodrugs **13c–d**, the carbamate derivatives of nitrophenyls and piperazine, were synthesized using Rivett and Wilshire's method. The synthesis begins with 1,4-bis(chlorocarbonyl)piperazine, obtained from the reaction of piperazine with phosgene. This intermediate is then reacted with nitrophenols (2-nitro, 3-nitro, and 4-nitro) in DMF at room temperature using NaH as the base [29].

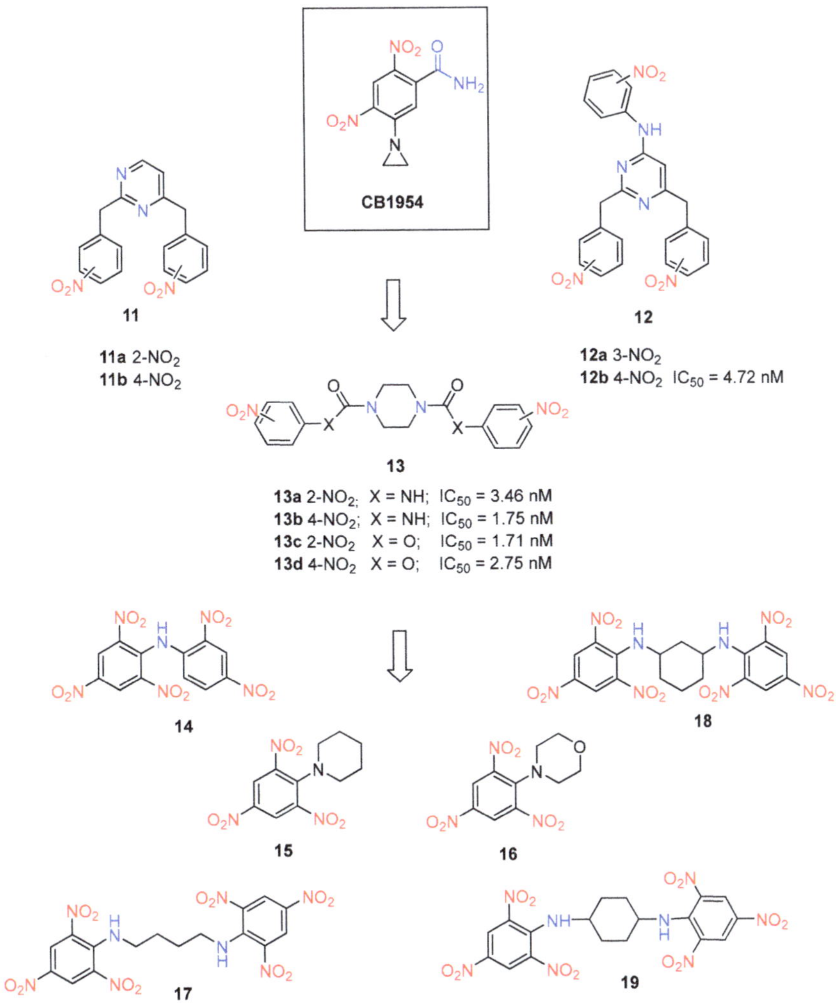

Figure 5. Structures of prodrug candidates **11–19** and a half-maximal inhibitory concentration (IC_{50}) of their metabolites.

The compounds displayed varying cytotoxic profiles. For example, the pyrimidine derivative **11b** and the triazine derivative **12a** emerged as promising drug candidates for prostate cancer with IC_{50} values of 54.75 μM and 48.9 μM, respectively. Compounds **12b**, **13a–c** were identified as prodrug candidates due to their non-toxic properties across three different cell models. The prodrug capabilities of these selected compounds were assessed using the SRB assay in combination with Ssap-NtrB. SRB screening results indicated that the metabolites of all selected non-toxic compounds exhibited significant cytotoxicity, with

IC_{50} values ranging from 1.71 to 4.72 nM, against prostate cancer cells. Among the tested compounds, piperazine derivatives **13b** and **13c** showed particularly notable toxic effects, with IC_{50} values of 1.75 nM and 1.71 nM, respectively, against PC3 cells, comparable to the standard prodrug CB1954 (IC_{50} = 1.71 nM) [29].

Based on enzymatic studies, prodrugs **15** and **18** (Figure 5) demonstrated the highest activity with Ssap-NtrB during short incubation periods, and their metabolite profiles were examined in detail over time. Similarly, **15** and **18** showed efficient enzymatic reduction by Ssap-NtrB. In contrast, **14** and **15** exhibited no interaction with nitroreductase within the limited timeframe, and the interaction levels for **17** and **19** were found to be insufficient. Furthermore, kinetic studies revealed that the catalytic efficiencies of the Ssap-NtrB/TNA1, Ssap-NtrB—**15**, and Ssap-NtrB—morpholine analog **16** combinations were 61, 28, and 20 times higher, respectively, than that of *E. coli* NfsB-CB1954 [30].

Although brief, this review highlights recent advances in the synthesis and structural modifications of nitroaromatic prodrugs, with potential applications in suicide gene therapy. It is hoped that these insights will pave the way for the design and synthesis of novel bioreductive agents for GDEPT.

3. Bioreductive-Activated Prodrugs Conjugates (BAPCs)

Another strategy for cancer therapy that targets hypoxia involves using hypoxia-activated prodrugs (triggers), which preferentially release chemotherapeutic agents (effectors) within hypoxic tumor regions. Under oxygen-poor conditions, the functional groups in these prodrugs (such as nitrophenyl, nitrobenzyl, or nitroheteroaryl triggers) are selectively reduced by reductases to electron-donating groups such as amine ($-NH_2$) or hydroxylamine (-NHOH), resulting in a dramatic change in the electron density of the aromatic moiety. The released electrons cause fragmentation of the linker and release the cytotoxic agent into the tumor, while leaving non-hypoxic cells undamaged (Figure 6) [6]. This approach can enhance therapeutic effectiveness compared to conventional chemotherapeutic treatments by concentrating the drugs within hypoxic tumor environments. At the same time, it reduces the side effects and toxicity associated with the systemic distribution of traditional drugs on normoxic cells [31].

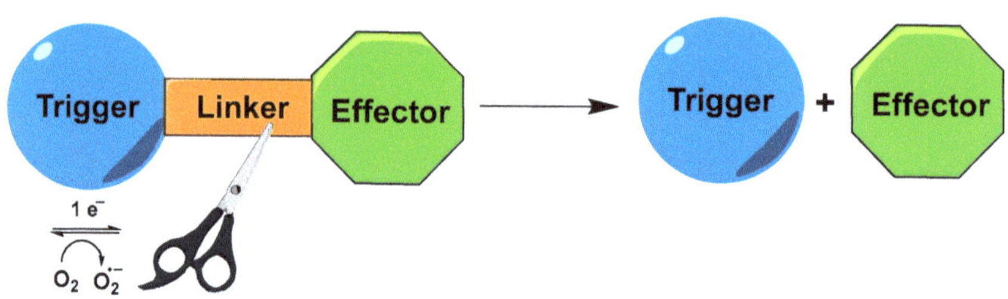

Figure 6. Reductive fragmentation of hypoxia-activated cytotoxins.

The trigger's role can be fulfilled by any of the above-mentioned bioreductive units, such as nitrophenyl, nitrobenzyl, nitro heteroaryl, azo compounds, quinones, or oxides [32–34]. The effector unit needs to have high cytotoxicity and effectiveness against multiple cancer types [35]. Commonly used effectors include drugs such as Doxorubicin (DOX), Camptothecin, and Paclitaxel (PTX) [36,37]. The linker, which connects the trigger to the chemotherapeutic agent, ensures stability in the bloodstream, while allowing for efficient release in the tumor environment. Preferred linkers are ether, ester, or carbamate subunits, due to their biocompatibility, stability in the bloodstream, and sensitivity to specific enzymes or acidic conditions, which are more prevalent in tumor tissues than in normal tissues [38].

By masking Fasudil's active site with a bioreductive 4-nitrobenzyl group, Al-Kilal et al. [39] synthesized the conjugate **20** (Figure 7). Under normoxic conditions, the conjugate exhibited significantly reduced antineoplastic activity (IC_{50} = 6.8 μM) compared to the parent compound (IC_{50} = 0.48 μM). However, under severe hypoxia, the nitro group is reduced to form an electron-donating substituent, which induces fragmentation and ejects the hydroxyfasudil **21.** This process significantly enhanced the antiproliferative effect on disease-afflicted pulmonary arterial smooth muscle cells and pulmonary arterial endothelial cells (IC_{50} = 0.40 μM).

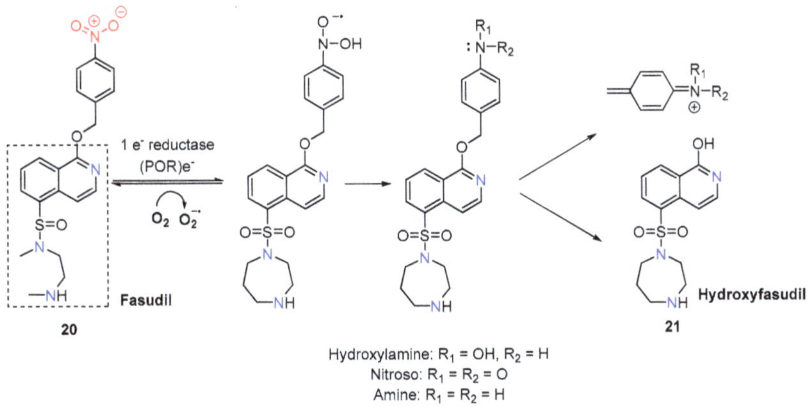

Figure 7. Scheme of the bioreduction process of conjugate **20** to hydroxyfasudil **21**.

Nitrobenzyl trigger was used in the construction of hypoxia-activated prodrug YC-Dox **22** (Figure 8) [31]. This prodrug is capable of specifically releasing the chemotherapeutic agent Dox and the HIF-1α (hypoxia-inducible factor-1α) inhibitor YC-1 hemisuccinate (3-(5′-hydroxymethyl-2′-furyl)-1-benzylindazole) in response to hypoxia.

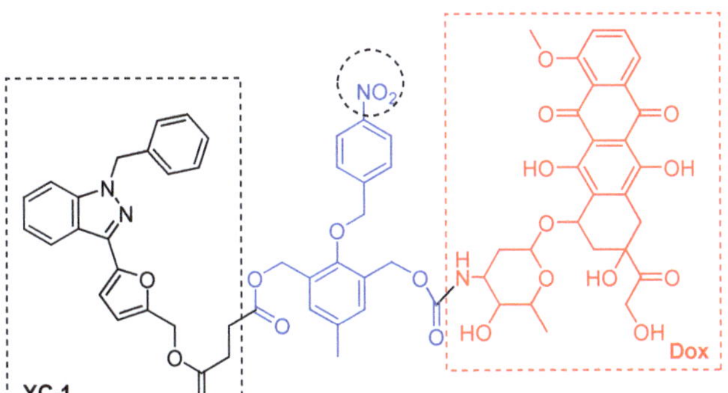

YC-Dox
22

Figure 8. Chemical structure of bioreductive-activated prodrug conjugate **YC-Dox 22**.

It is well known that low oxygen levels in hypoxic tumor tissues lead to the accumulation of HIF-1α. This protein plays a crucial role in the adaptive response of cancer cells to hypoxia by regulating various cellular functions [40]. YC-1 is capable of blocking HIF-1α expression and consequently inhibiting the activity of HIF-1 as a transcription factor in hypoxic cancer cells, leading to the suppression of tumor growth. In addition, YC1 exhibits antiproliferative effects [41,42]. The release of Dox and YC-1 from the prodrug YC-Dox in response to hypoxia results in substantial synergistic potency against hypoxic cancer cells and remarkable cytotoxic selectivity, being more than eight times greater compared to normoxic healthy cells. In vivo experiments demonstrate that this prodrug can selectively target hypoxic cancer cells while avoiding unintended effects on normal cells. This selective targeting results in enhanced therapeutic efficacy for tumor treatment and reduced adverse effects on normal tissues.

Ce, Y et al. reported the synthesis of the hypoxia-activated prodrug, *N*-(2-chloroethyl)-*N*-2-(2-(4-nitrobenzylcarbamate)-O^6-benzyl-9-guanine)ethyl-*N*-nitrosourea (NBGNU), **23** (Figure 9) [43]. The molecule integrates the chloroethylnitrosourea (CENU) pharmacophore to induce DNA interstrand cross-links and an O^6-benzylguanine analog moiety (angiotensinogen (AGT) inhibitor) masked by a 4-nitrobenzylcarbamate group to induce hypoxia-activated inhibition of O^6-alkylguanine-DNA alkyltransferase. Its anticancer effectiveness was assessed through in vitro experiments. The prodrug demonstrated promising antitumor efficacy and hypoxic selectivity due to the incorporation of an AGT inhibitor and hypoxia-activated pharmacophores into the side chain of the CENU moiety. The activity of **23** against AGT-expressing human glioma SF763 cells under hypoxic conditions (IC_{50} = 126 μM) was significantly higher than under normoxic conditions (IC_{50} = 580 μM). This indicates that NBGNU selectively undergoes reduction under hypoxic conditions, leading to the unmasking of the 2-amino group of guanine, the release of O^6-BG analogs, and the effective inhibition of AGT. However, to further enhance its hypoxic selectivity and chemotherapeutic efficacy, improvements are needed to reduce normoxia activation and increase water solubility.

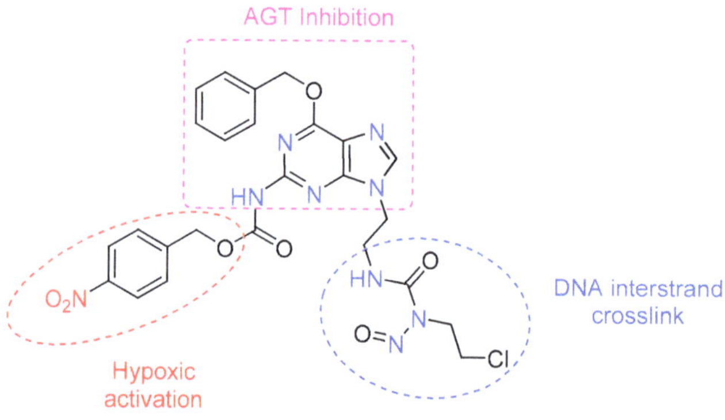

NBGNU
23

Figure 9. Chemical structure of bioreductive-activated prodrug conjugate **NBGNU 23.**

The 2-nitrobenzyl- and 4-nitrobenzyl-SN-38 analogs **24a**,**b** (Figure 10), as discussed by Liang et al. [44], embody a key feature of hypoxia-activated prodrugs: they are significantly less potent than their active metabolite. Initial cell viability assays showed that these analogs were considerably less cytotoxic than SN-38 against human leukemia K562 cells, with 2-nitrobenzyl-SN-38 **24a** and 4-nitrobenzyl-SN-38 **24b** displaying 8-fold and 19-fold lower cytotoxicity, respectively. Furthermore, in a topoisomerase I assay, the 4-nitrobenzyl analog at the C-10 position of SN-38 inhibited the enzyme's ability to relax supercoiled pBR322 DNA at concentrations similar to the clinically approved SN-38. Although the reduction potentials of these compounds were lower than those of other known HAPs and partially reversible, they demonstrated potential as hypoxia-targeted therapeutics. The study concluded that the next generation of SN-38-HAPs should incorporate bulkier nitroaromatic groups to further reduce cytotoxicity and use triggers with higher reduction potentials to align with the range of cellular reductases (−450 to −300 mV).

The 2-nitroimidazole fragment is a widely used trigger in the fragmentation concept, due to its good hydrophilicity and its relatively high one-electron reduction potential, well within the range of various reductase enzymes, and high selectivity for hypoxic conditions. Furthermore, there is a significant body of clinical and preclinical data supporting the efficacy and safety of 2-nitroimidazole-based prodrugs. This accumulated evidence provides a strong foundation for their continued use and further development.

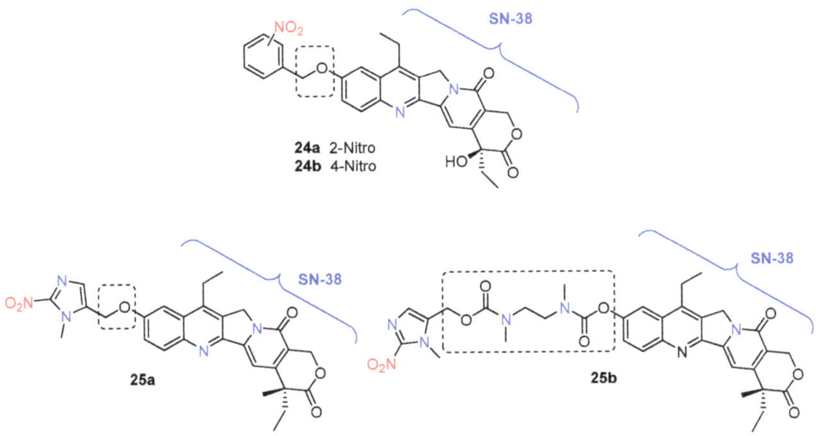

Figure 10. Chemical structure of bioreductive-activated prodrugs conjugates **24a,b** and **25a,b**.

Choi and co-workers [45] conjugate the camptothecin derivate SN-38 with 1-methyl-2-nitro-1*H*-imidazole-5-yl fragment using two different linkers—an ether linkage **25a** and carbamate functionality **25b** (Figure 10). The ether analog **25a** had moderate hypoxia selectivity and more toxicity compared with TH-302. The different linkers in the structures of the two derivatives likely account for the significant differences in their hypoxia selectivity and toxicity. Compound **25a** demonstrated ten times higher toxicity against the human lung cancer cell line H460 and the human colon cancer cell line HT29 compared to the control hypoxia-activated nitroimidazole prodrug TH-302. Furthermore, **25a** exhibited lower toxicity than SN-38 under normoxic conditions. However, both the hypoxic selectivity and toxicity of **25a** were lower compared to those of compound **25b**. Despite this, the ether-linked compound 3a is considered a promising hypoxia-selective antitumor agent.

Bielec, B. and colleagues [46] developed the first crizotinib prodrugs **26a,b** (Figure 11) aimed at reducing severe adverse effects and enhancing anticancer activity. The design of these prodrugs involves a hypoxia-activatable, self-immolative 2- nitroimidazole trigger moiety at a key tyrosine kinase binding site of crizotinib, which significantly reduces its affinity for the catalytic pockets of the target kinases c-MET and ALK. Two different prodrug derivatives were synthesized: one with the trigger moiety coupled via carbamoylation (**26a**) and the other via alkylation (**26b**) of the 2-aminopyridine moiety of crizotinib. Prodrug **26a** demonstrated high stability in serum, a crucial requirement for successful prodrug development, and effectively inhibited c-MET phosphorylation and cell proliferation in tumor tissues in vivo following intravenous application. Overall, the data suggest that prodrug **26a** is a promising candidate for further (pre)clinical development as a novel tyrosine kinase inhibitor with improved tumor-specific properties.

In another study, a hypoxia-activated camptothecin derivative embodies a multifunctional bioreductive linker based on 1-methyl-2-nitroimidazole. The incorporation of a PEG chain in the linker increased the water solubility of the SN-38- prodrug **27** and ensured stability under physiological conditions [47]. When conjugated with SN-38, this linker demonstrated the capability to efficiently release the drug through a two-step process: reductive activation of the 2-nitroimidazole, followed by spontaneous degradation of the linker via 1,6-elimination and cyclization-elimination, ultimately resulting in drug molecule release (Figure 12).

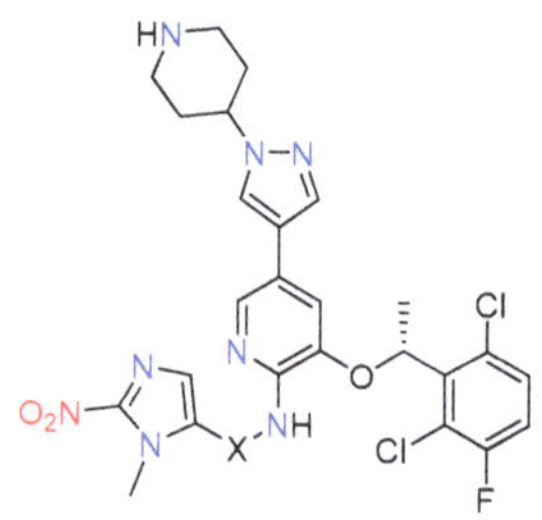

26a: X = OCO
26b: X = CH_2

Figure 11. Chemical structure of bioreductive-activated prodrugs conjugates **26a**,**b**.

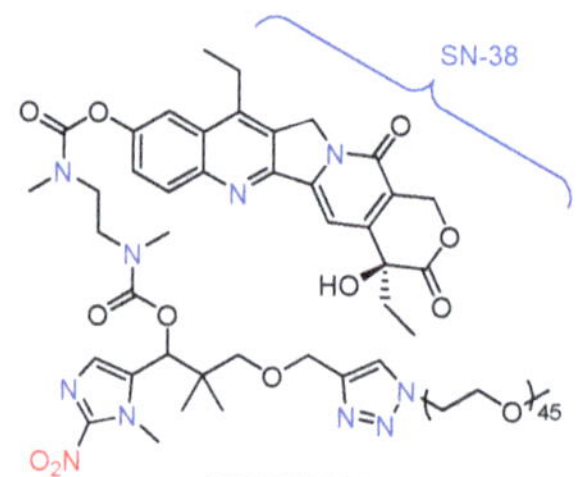

I5S-PEG2000
27

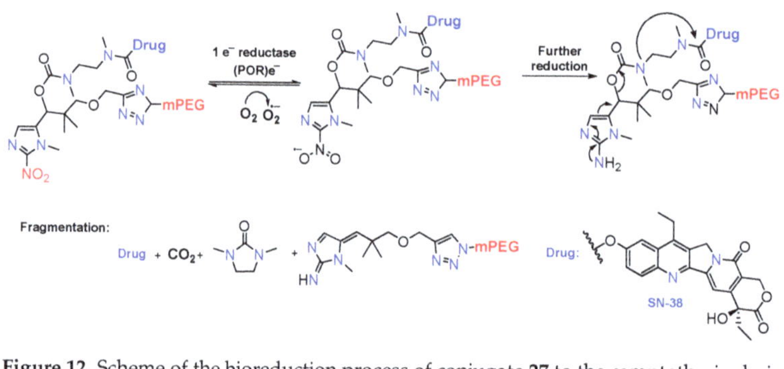

Figure 12. Scheme of the bioreduction process of conjugate **27** to the camptothecin derivative SN-38.

Encouraged by the great properties of the multifunctional linger-containing 1-methyl-2-nitroimidazole trigger unit [47], Chang et al. reported the synthesis and biological evaluation of the hypoxia-activated albumin-binding prodrug Mal-azo-Exatecan **28** (Figure 13) [48]. The 5-position branched linker of 1-methyl-2-nitro-5-hydroxymethylimidazole served as a hypoxic cleavage trigger, linking the camptothecin analog Exatecan via a carbamate bond. After intravenous administration, the side-chain maleimide rapidly binds to human serum albumin (HSA). The HSA-azo-Exatecan carrier system accumulates in tumor tissue through the enhanced permeability and retention effect, as well as the interaction with the albumin receptor gp60. In the hypoxic tumor environment, Exatecan is released, triggered by nitroreductase. The nitroimidazole trigger has high plasma stability and does not cause the chemotherapeutic agent release from HSA-azo-Exatecan during circulation in vivo, avoiding systemic side drug effects.

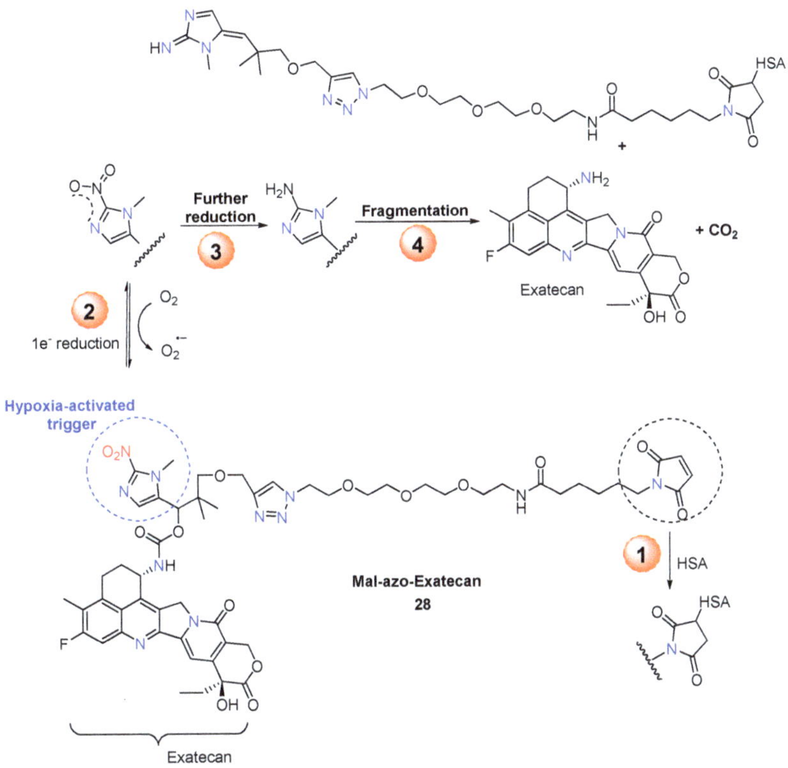

Figure 13. Scheme on the bioreduction process of conjugate **28** to Exatecan.

Zhang and co-workers [49] reported the synthesis of hypoxia-activated paclitaxel (PTX) prodrugs IMI-PXT **29** based on hypoxia-sensitive 2-nitroimidazole moiety. The 2-nitroimidazole unit at the N-1 position is connected to 2′-OH of PTX via an ester bond with pivalic acid for easier drug release. The GLU-PTX **30** and AZO-PTX **31** prodrugs (Figure 14) contain glucose and acetazolamide as targeting ligands.

In contrast to the 2-nitroimidazole unit, 2-nitrothiophene and 2-nitrofuran triggers in hypoxia-activated prodrugs exhibit less favorable stability, reactivity, and toxicity profiles, rendering them less attractive for the development of HAPs.

Winn et al. reported the synthesis of the scombretastatin A-1 (CA1) and combretastatin A-4 (CA4) prodrug conjugates **32a,b** (Figure 15) [50]. The most active compounds in the series were the gem-dimethyl prodrugs of CA1 (**32a**) and CA4 (**32b**), exhibiting hypoxia cytotoxicity ratios of 12.5 and 41.5, respectively. This high selectivity is attributed to the gem-dimethyl CA4-BAPC's enhanced resistance to cleavage in oxygenated environments, allowing the parent anticancer agent (CA4) to be released selectively under hypoxic conditions.

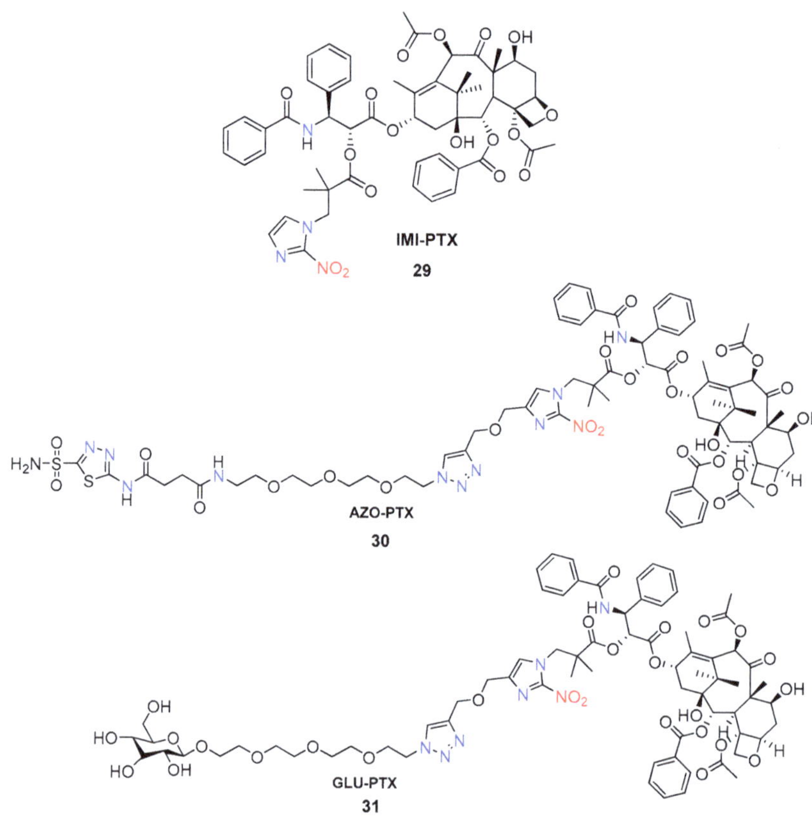

Figure 14. Chemical structure of bioreductive-activated prodrugs conjugates **29–31**.

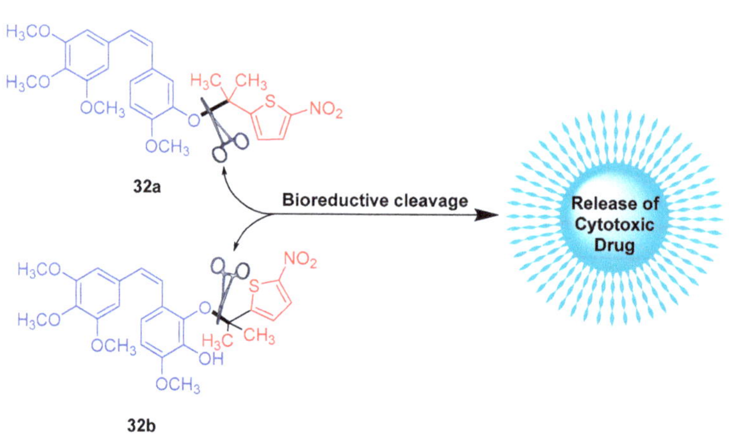

Figure 15. Chemical structure of gem-dimethyl prodrugs **32a,b**.

4. Nitroaromatic Compounds as Fluorescent Probes for Hypoxia Detection and Imagining

Conventional methods for the in vivo imaging and detection of solid tumors are typically applied only in advanced stages of cancer. Hypoxia imaging, however, offers a promising alternative for earlier cancer diagnosis, enabling tumor visualization with a diameter as small as 350 μm [51]. The major techniques for hypoxia measurements in tumors include immunohistochemical staining, oxygen electrodes, DNA strand breaks, polarographic needle electrodes, magnetic resonance imaging, positron emission tomography, single-photon emission computed tomography (SPECT), and fluorescence imaging. In recent decades, fluorescence imaging has emerged as one of the most advanced methods for quantifying hypoxia. It offers several advantages, such as non-invasiveness, higher sensitivity, real-time monitoring in living systems, absence of ionizing radiation, low toxicity, simple operation, and low cost [51–55].

Due to the relatively easy synthesis, well-predictable and highly selective fluorescent sensing output, the nitro aromatic molecules have become the most attractive approach in the design of fluorescent probes for hypoxia conditions [56–59]. Even commercially available options for in vivo studies of hypoxia, such as pimonidazole (alpha-((2-Nitroimidazol-1-yl)methyl)-1-piperidineethanol), are based on nitro-containing compounds [60].

To date, the nitro aromatic compounds were used as a platform for fluorescent recognition of hypoxia according to two major mechanisms, and both were based on the bioreduction of the nitro aromatic system. The first one referred to the selective labeling of hypoxic cells due to the reduction of a nitroaromatic moiety in the fluorophore architecture to amine. The main concept here lies in the fact that the nitro group is well known as a fluorescence quencher in aromatic systems, but after the bioreduction of this nitro group in hypoxic conditions, the aromatic systems become fluorescent. The second mechanism was based on the reduction of a fluorescent probe containing 4-nitrobenzyl formiatic, 4-nitrobenzylic, or similar heterocyclic fragments in hypoxic cells, which resulted in a scavenge reaction and alterations in former fluorescence wavelength or intensity (Figure 16).

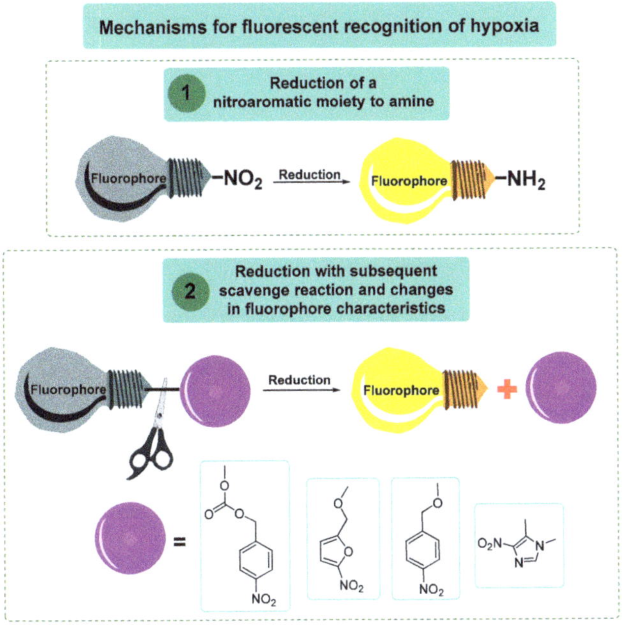

Figure 16. The basic approaches for designing fluorescent hypoxia-activated probes.

The development of fluorescent probes for hypoxia imaging began in the 1980s and 1990s, focusing on the bioreduction of nitroaromatic compounds to amines. Olive and Durand were among the first to reveal the significant potential of nitroaromatic rings for hypoxia imaging [61,62]. In their reports, they demonstrated that relatively nontoxic nitrofurans **33–36** (Figure 17) exhibit a highly responsive fluorescent output to intracellular oxygen concentration under hypoxic conditions.

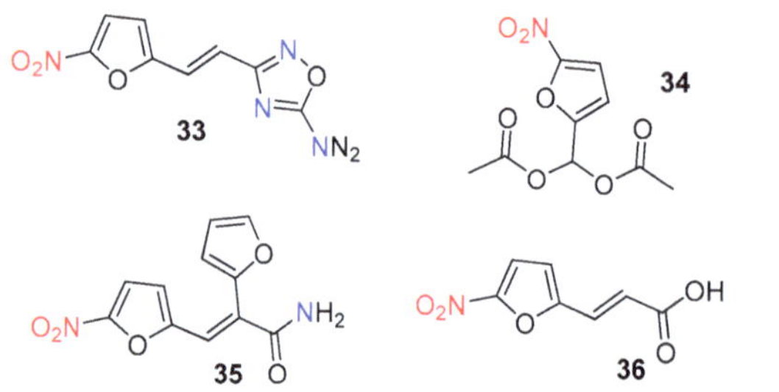

Figure 17. Chemical structures of hypoxia-activated fluorescent probes **33–36**.

Later, Hodgkiss et al. obtained a family of hypoxia-activated fluorescent naphthalimides (**37–42**) containing 2-nitroimidazole side chains (Figure 18). These molecules were nonfluorescent due to the photoinduced electron transfer (from the fluorophore-excited state to the nitroaromatic moiety), which was prevented after enzymatic bioreduction [55]. Furthermore, it is well known that the bioreduction of nitroimidazoles leads to intracellular bioreductive metabolites that react with biomolecules, thus providing a binding mechanism for the fluorescent probes to hypoxic cells. Since this report, the 2-nitroimidazole side chains have become a major tracer for hypoxic cells and remain popular to date.

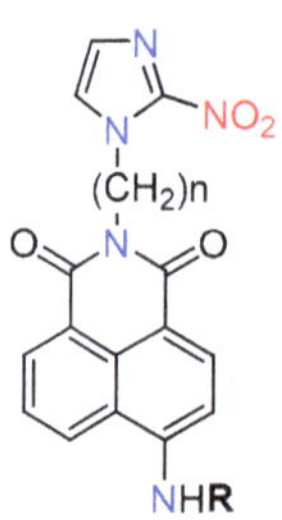

37 R = H, n = 3

38 R = H, n = 5

39 R = Ac, n =3

40 R = $COCH_2Cl$, n = 3

41 R = $COCH_2NH(CH_2)_2N(CH_2CH_2OH)_2$, n = 3

42 R = H, n = 9

Figure 18. Chemical structures of hypoxia-activated 1,8-naphthalimide probes **37–42**.

Due to the relatively easy synthesis and well-predicted off-on fluorescent response to intracellular oxygen concentration, numerous nitroaromatic structures have been developed for hypoxia imaging based on the two principles mentioned above: the bioreduction of a nitro group directly attached to the fluorophoric system, or the use of 2-nitroimidazole side chains. For example, Qian et al. extended the concept of Hodgkiss et al. by preparing similar naphthalimides (**43–46**, Figure 19) but with two 2-nitroimidazole fragments instead of one, thus reporting the first hypoxic probes containing two heterocyclic-binding side chains [63]. The main motive for the synthesis of compounds **44** and **46** was the study of the side chain effect in C4-position of the naphthalimide ring, which plays a more important role than the fluorophoric architecture itself during interaction with DNA and probe interference. The authors discovered that in V79 cells, probe **46** exhibited a higher fluorescence enhancement (FE = 20 times) compared to **44** (FE = 15 times). This difference in signal responses between the two probes was attributed to the shorter side-chain length of **44** compared to **46,** which reduces the possibility of fluorescent quenching due to intramolecular photoinduced electron transfer from the reduction products of the nitroimidazole moiety to the excited naphthalimide fluorophore.

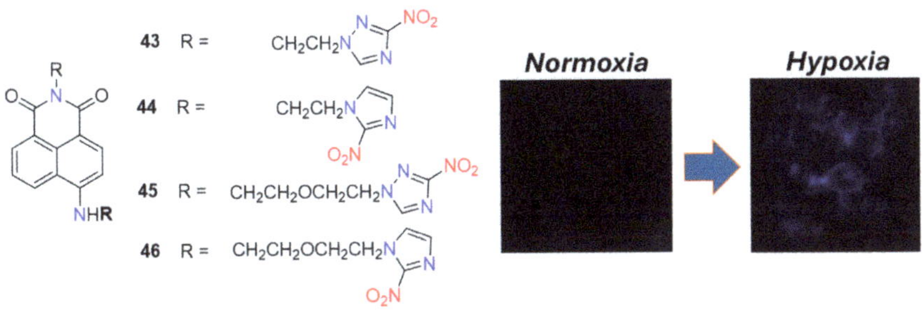

Figure 19. Chemical structures of hypoxia-activated 1,8-naphthalimide probes **43–46** and V79 cells incubated with **46** under normoxic and hypoxic conditions. Adapted with permission from [63]. Copyright (2006) Elsevier.

Qian et al. were motivated by the ease of synthesis and low cost to develop a series of hypoxic probes (**47–53**) containing a nitro group directly incorporated into the fluorophoric scaffold (Figure 20) [64,65]. Compounds **47–49** demonstrated promising fluorescent responses in hypoxic V79 cells, with the hypoxic-oxic fluorescence differential reaching 6, 9, and 11 times after incubation with **47**, **48**, and **49**, respectively. However, the observed increase in fluorescence enhancement correlated with the probes' increased water solubility. This indicates that water solubility was a general issue with these probes, as deposition outside the cells hindered future quantitative analysis.

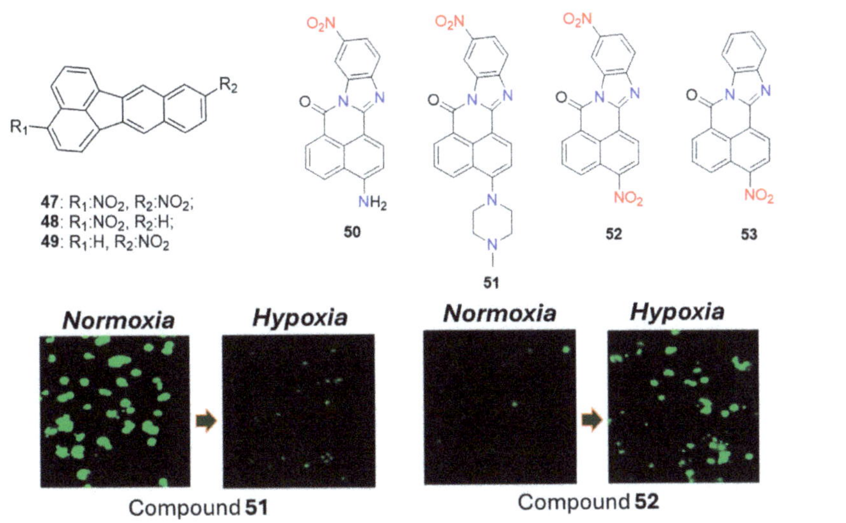

Figure 20. Chemical structures of hypoxia-activated 1,8-naphthalimide probes **47–53** and V79 cells incubated with **51** (left) and **52** (right) under normoxic and hypoxic conditions. Adapted with permission from [65]. Copyright (2008) Springer Nature.

Furthermore, with probes **50–53**, Qian et al. demonstrated that the NO_2 group in fluorogenic compounds could function not only as a fluorescent quencher but also as an electron acceptor that enhances fluorescence emission. Probes **52** and **53** exhibited the usual off-on fluorescent switching upon transitioning from oxic to hypoxic conditions, showing a 12-fold fluorescence enhancement. However, the analogous probes **50** and **51**, which contained the nitro group only in the electron-accepting part of the fluorophoric system, displayed the opposite fluorescent response—a 14-fold fluorescence quenching under hypoxic conditions compared to oxic conditions. This was attributed to the internal charge transfer (ICT) nature of the fluorophore, where fluorescence appears due to the charge transfer from electron-donating amines to the nitro-containing electron-accepting part of the molecule. The bioreduction of this nitro group to an electron-rich amine destabilized the ICT state, which is crucial for strong fluorescence. Additionally, the authors found that under hypoxic conditions in V79 cells, the nitro group in the naphthalene ring was reduced preferentially over the one in the benzene ring. This selective reduction was essential for maintaining the observed fluorescent on-state of probe **52** in hypoxia.

The concept of incorporating a nitro group directly into the fluorophoric system, which activates fluorescence emission due to the favored intramolecular charge transfer after bioreduction, remains relevant for designing hypoxic probes. Recently, Fan et al. reported a non-fluorescent benzothiazole probe, **54** (Figure 21) [66]. Upon reaction with nitroreductase (NTR), the nitro group in probe 54 underwent enzyme-catalyzed reduction to an amine, resulting in a strong fluorescent emission. This process demonstrated high sensitivity, with a detection limit of 48 ng/mL and a linear range of 0.5–8.0 μM for NTR. Additionally, **54** was successfully used for imaging hypoxia levels in living HeLa cells, rat tumor tissues, and zebrafish. These results, combined with the probe's low toxicity, indicate significant potential for detecting hypoxia in solid tumors.

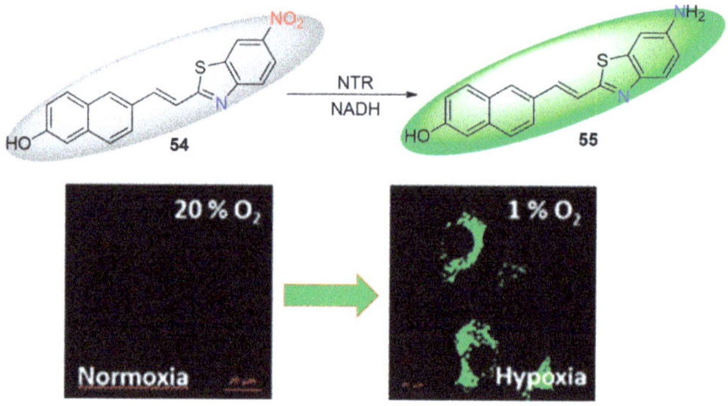

Figure 21. Chemical structures of hypoxia-activated probe **54** and HeLa cells incubated with **54** under normoxic and hypoxic conditions. Reproduced with permission from [66]. Copyright (2020) The Royal Society of Chemistry.

Janczy-Cempa et al. employed a similar strategy in the design of two nitro-pyrazinotriazapentalene derivatives, **56** and **57** (Figure 22), initially exhibiting weak fluorescence [67]. Upon reduction of their nitro groups by NTR, a significant fluorescence enhancement with a 15-fold increase in intensity was observed. The reduction process with NTR was selective, with linear ranges of 0–4 μg/mL and limits of detection of 18.6 ng/mL for **56** and 33.2 ng/mL for **57**, respectively. Both probes were non-toxic and successfully employed for imaging hypoxia in the A2058 cell line.

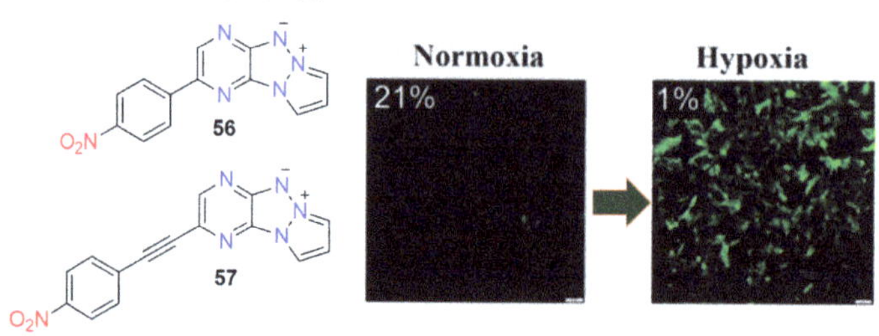

Figure 22. Chemical structures of hypoxia-activated 1,8-naphthalimide probes **56** and **57**, and HeLa cells incubated with **56** under normoxic and hypoxic conditions. Adapted with permission from [67]. Copyright (2021) Elsevier.

The current focus on developing highly emissive fluorescent probes in the near-infrared (NIR) region is driven by NIR's superior penetration ability in living systems, effectively minimizing interfering bio-autofluorescence. This motivation led Fan et al. to synthesize a BODIPY-based fluorescent probe, **58**, for hypoxia imaging (Figure 23) [68]. Initially non-emissive, compound **58** contained a fluorescent-quenching nitro group that could be selectively reduced by NTR to its amine form. Upon reduction, the resulting amino derivative exhibited bright fluorescence in the NIR spectrum at 713 nm. Unlike its precursor, the fluorescence of the amino derivative of **58** showed a linear increase, correlated with the NTR concentration in the range of 0.1−1.0 μg/mL. A significant 55-fold maximum fluorescence enhancement was achieved, with a calculated limit of detection (LOD) of 7.08 ng/mL. Building upon probe **58**, nanoparticles were developed and used as fluorescent probes for hypoxia imaging both in vitro (H9c2 cells) and in vivo (ischemic mice model). These nano-probes demonstrated negligible toxic effects, making them suitable for in vivo applications.

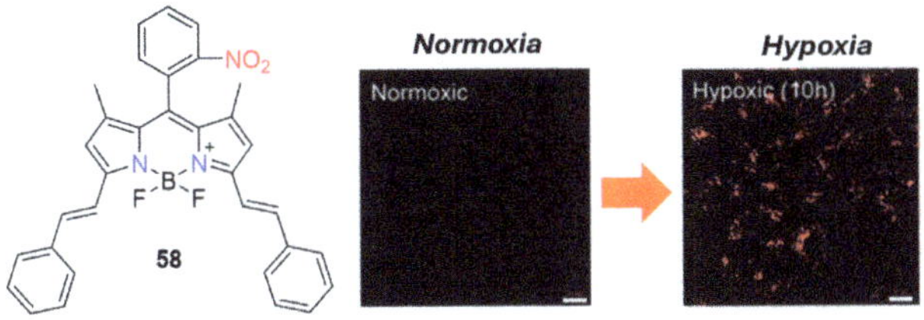

Figure 23. Chemical structures of hypoxia-activated probe **58** and H9c2 cells incubated with **58** under normoxic and hypoxic conditions. Adapted with permission from [68]. Copyright (2019) American Chemical Society.

Probe **59** (Figure 24) represents another notable example of a NIR fluorescent probe designed for hypoxia imaging, featuring a directly attached nitro group as a recognition unit for NTR [69]. Upon selective reduction by NTR, probe **59** exhibited remarkable fluorescence, peaking at 740 nm. This resulted in a 32-fold enhancement in fluorescence intensity and an exceptionally low detection limit of 1.09 ng/mL. Probe **59** was effectively utilized for visualizing hypoxic cancer cells (HeLa, HepG2) and hypoxic tumors in a tumor-bearing mouse model. Furthermore, MTT analysis indicated negligible cytotoxicity, underscoring its potential as a highly valuable tool for both in vitro and in vivo monitoring of hypoxia status.

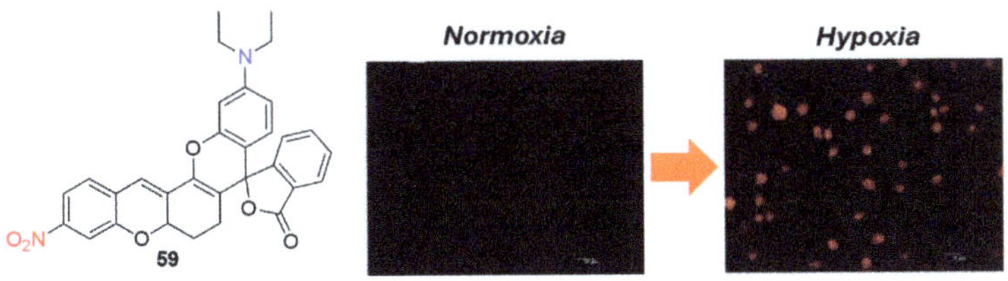

Figure 24. Chemical structures of hypoxia-activated probe **59** and HeLa cells incubated with **59** under normoxic and hypoxic conditions. Adapted with permission from [69]. Copyright (2022) Elsevier.

The integration of nitroimidazole side chains into NIR fluorophores represents another effective approach for designing turn-on NIR fluorescent hypoxia probes. A notable example includes probes **60–62** (Figure 25), where the fluorophore system is linked to two fluorescence-quenching nitroimidazole chains [70]. These probes are designed to preferentially accumulate under hypoxic conditions and exhibit strong fluorescence in the range of 700–900 nm, following the bioreduction of both nitroimidazole fragments. Probes

60–62 have been successfully applied for imaging hypoxic tumors both in vivo and in vitro, highlighting their potential utility in hypoxia research and cancer diagnostics.

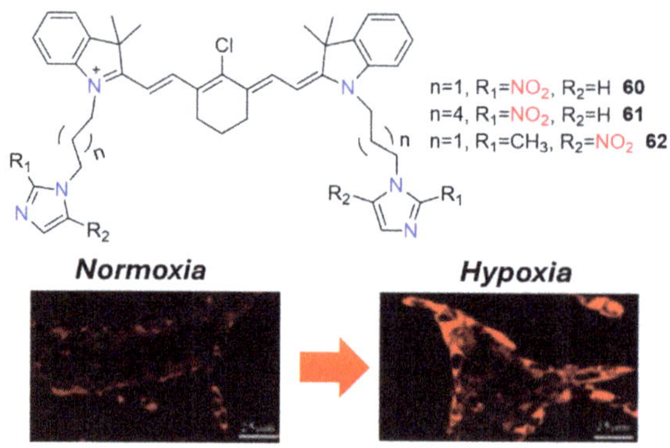

Figure 25. Chemical structures of hypoxia-activated probes **60–62** and 4T1 cells incubated with **61** under normoxic and hypoxic conditions. Adapted with permission from [70]. Copyright (2021) American Chemical Society.

The second main approach for designing hypoxic fluorescent probes involves fluorogenic systems that incorporate 4-nitrobenzyl formate, 4-nitrobenzylic, or similar heterocyclic fragments into hypoxic cells, typically binding to amino or hydroxy groups. This approach offers greater flexibility in designing signaling activation compared to the simple reduction of aromatic nitro groups to amines discussed earlier. These probes undergo a scavenging reaction during bioreduction, allowing for a variety of photophysical signaling mechanisms such as ICT, PET (photoinduced electron transfer), and ESIPT (excited-state intramolecular proton transfer). Compound **63** (Figure 26) serves as a classic example of a probe containing 4-nitrobenzyl formate, designed for detecting NTR [71]. It features a rhodolite fluorophore system known for its high quantum yield and stability across a wide pH range. In **63**, the presence of nitrobenzyl formate quenches the rhodol fluorescence, but this quenching is easily reversed in the presence of NTR. Upon reduction, a strong fluorescence emission is observed, facilitating the selective determination of NTR. The fluorescence intensity at 550 nm increases approximately 4.3-fold upon activation, with a reported detection limit of 51.5 ng/mL for NTR. In studies involving Hi-5 cells, compound **63** exhibited non-toxic effects and good cell permeability. It was successfully utilized for imaging hypoxic conditions both in vitro, using Hi-5 cells as a model, and in vivo, in *C. elegans*.

Probe **65,** as reported by Wei et al. (Figure 27), is based on a naphthalimide fluorogenic architecture and serves as another example of the efficient use of the 4-nitrobenzyl formiatic-recognizing unit in the detection of hypoxia [72]. The fluorescence intensity of **65** was increased with the increased concentration of NTR, with a linear range of 0.1–0.3 μg/mL. From the observed standard plot, a detection limit of 0.1 μM was calculated. Furthermore, this probe showed remarkably low cytotoxicity, as even at high concentrations of **65,** the percentage of cell viability remained above 95%. The confocal fluorescence imaging of U87 cells revealed the great potential of the probe to monitor the hypoxic status of tumor cells.

The tumor cells were characterized not only by a hypoxic environment but also by increased acidity. That is why the simultaneous detection of acidity and NTR could reduce the possibility of false positive results during intracellular tumor imagining. From this point of view, compound **67** (Figure 28) represents a noteworthy probe for accurate tumor imaging due to its capability to detect both acidity and NTR [73]. In probe **67** a 4-nitrobenzyl formiatic fragment was introduced in a classic PET (photoinduced electron transfer) 4-amino-1,8-naphthalimde pH sensor based on a "fluorophore-spacer-receptor" model where the 4-amino-1,8-naphthalimide serves as the fluorophore, and morpholine acts as the pH receptor. After excitation, electron transfer from the electron-rich morpholine amine to the fluorophore occurs in this molecule, which quenches the fluorescence emission. Upon protonation in acid media, the morpholine formed an electron-poor quaternary ammonium salt, the PET process became impossible, and blue fluorescence was registered. However, the observed fluorescence was weak due to the presence of hypoxia-recognition 4-nitrobenzyl formiatic unit, which quenches emission too. In neutral hypoxic media, the 4-nitrobenzyl formiat in **67** was reduced selectively to form a green-emitting compound. As a result of the reduction-scavenging reaction, the strong electron-accepting carbonyl coupled to the 4-amino nitrogen in 1,8-naphthalimide was cut off. This led to an increase in the electron-donating ability of the 4-amino substituent attached to the 1,8-naphthalimide, thus increasing the fluorophore ICT efficiency and causing the fluorescence emission to red-shift from the blue to the green region. The observed green emission was weak too due to the PET quenching effect in neutral media mentioned above. When the probe is in an acid and hypoxia environment, both quenching processes are blocked, and bright green fluorescence appears. Furthermore, based on both fluorescence outputs (blue at 460 nm and green at 524 nm), a ratiometric analysis was conducted. In ratiometric methods for analyte determination, the quantification is based on the ratio of fluorescent intensities at two different wavelengths. This approach is desirable in bioimaging because it allows for self-calibration and built-in correction for environmental effects and biomolecules. The pH 5 ratiometric analysis for the detection of NTR showed linearity in the range of 0–20 μM and a limit of detection of 0.92 μg/mL. The probe was applied for the fluorescence imaging of acidity and hypoxia in A549 cells.

Zheng et al. elegantly demonstrated the use of 4-nitrobenzyl formiate as an NTR recognition unit combined with NIR imaging for in vivo hypoxia detection, as shown in their cyanine probe **68** (Figure 29) [74]. The fluorescent analysis of **68** revealed a linear enhancement at 785 nm in the presence of 0–0.5 μg/mL NTR, with a low detection limit (LOD) of 0.0242 μg/mL. The efficient NTR detection capability of **68** was successfully applied for the fluorescence imaging of hypoxic A549, PC-12, and HUVEC cell lines. Furthermore, in vivo hypoxia imaging using probe **68** was evaluated in tumor-bearing mice, as well as in models of cerebral ischemia and deep vein thrombosis. The results underscored the high potential of probe **68** for rapid and precise in vivo monitoring of NTR activity across diverse clinical models.

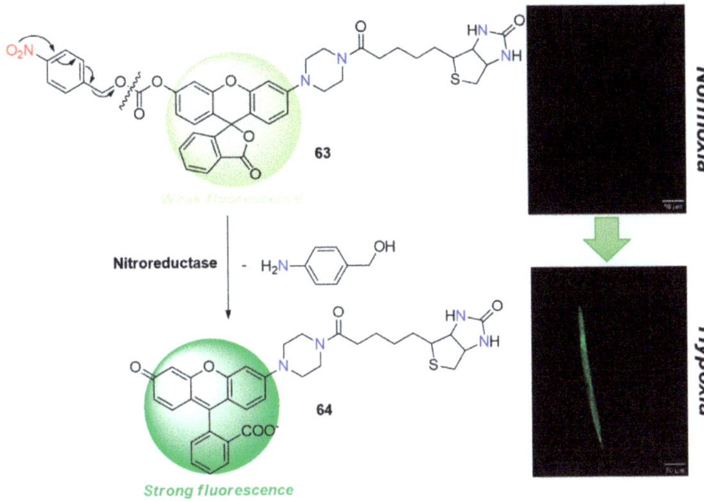

Figure 26. Chemical structures of hypoxia-activated probe **63** and C. elegans incubated with **63** under normoxic and hypoxic conditions. Reproduced with permission from [71]. Copyright (2017) The Royal Society of Chemistry.

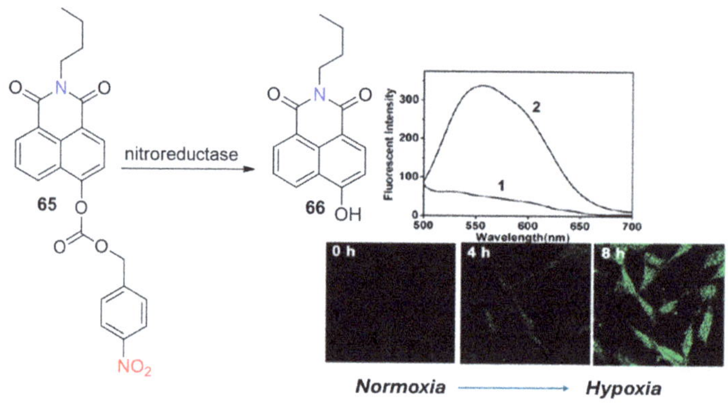

Figure 27. Chemical structures of hypoxia-activated probes **65** and U87 cells incubated with **65** under normoxic (1) and hypoxic (2) conditions. Adapted with permission from [72]. Copyright (2018) Elsevier.

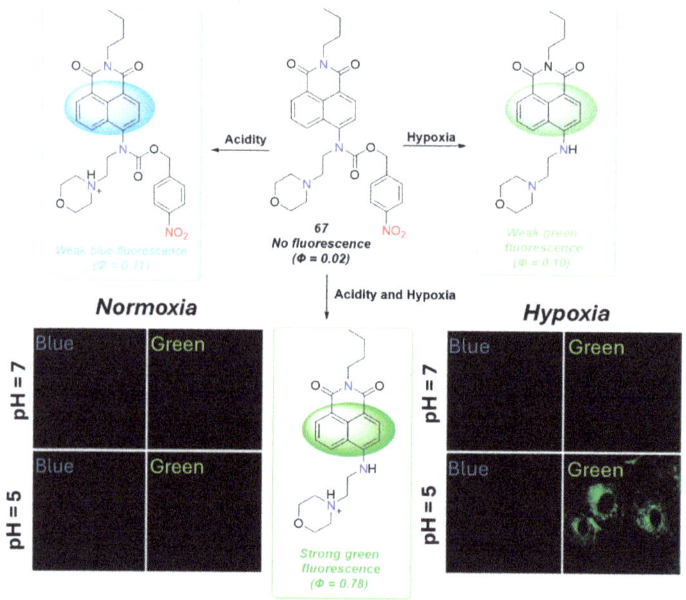

Figure 28. Fluorescence sensing mechanism of probe **67** for detection of acidity and hypoxia, and A549 cells incubated with **67** under normoxic and hypoxic conditions, at different pHs. Adapted with permission from [73]. Copyright (2018) The Royal Society of Chemistry.

Using 4-nitrobenzyl instead of the 4-nitrobenzyl formiate fragment as a recognition unit offers another approach in designing selective probes for detecting NTR and imaging hypoxic cells. Probe **69** (Figure 30) exemplifies this strategy, where the reduction by nitroreductase leads to the formation of a green-emitting fluorescent compound through the scavenging of the 4-nitrobenzylic unit and intramolecular cyclization [75]. Due to the specific interaction of the 4-nitrobenzylic moiety with NTR, compound **69** exhibits increased fluorescence intensity at 530 nm, with a linear response in the concentration range of NTR from 0 to 10 μg/mL and a detection limit of 11 ng/mL. The probe has been successfully used to visualize hypoxic conditions in living HepG2 cells, demonstrating low toxicity and the ability to detect NTR in tumor tissues up to a depth of 100 μm.

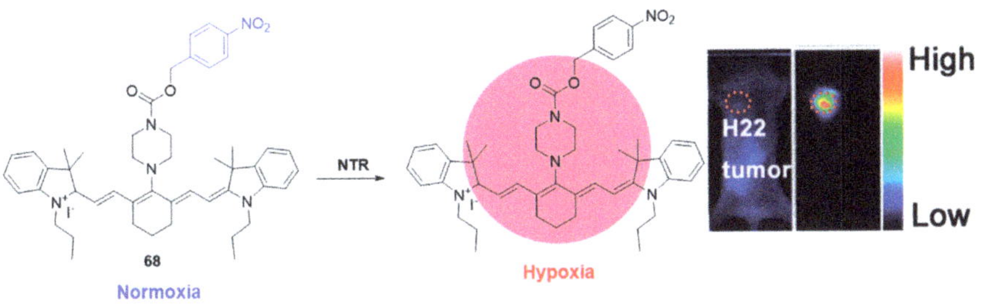

Figure 29. Fluorescence sensing mechanism of probe **68** for detection of hypoxia, and in vivo hypoxia-activated tumor imagining with **68**. Adapted with permission from [74]. Copyright (2018) Elsevier.

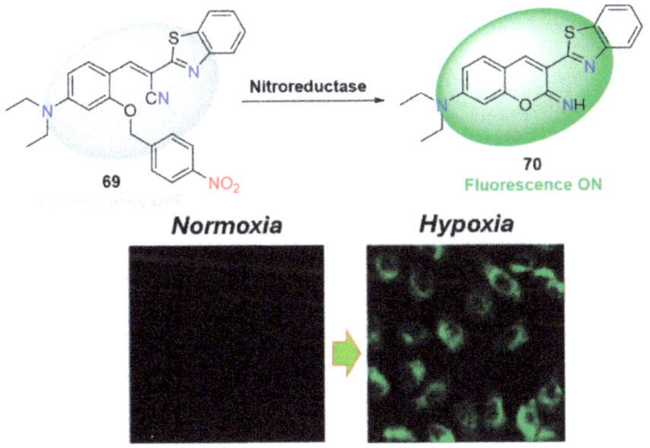

Figure 30. Fluorescence sensing mechanism of probe **69** for detection of NTR and HepG2 cells incubated with **69** under normoxic and hypoxic conditions. Adapted with permission from [75]. Copyright (2018) Elsevier.

Figures 31 and 32 illustrate two fluorescent probes (**71** and **72**) as typical examples for in vivo NIR imagining of hypoxic conditions using the NTR-selective 4-nitrobenzyl recognition unit [76,77]. Compound **71** represents a high-efficiency NIR fluorescence probe for the detection of hypoxia via responding to NTR. It showed gradually increased fluorescence at 710 nm in the presence of NTR. The observed LOD was 13.441 ng/mL within the linear range of 0.1–0.9 μg/mL. The in vitro confocal-mediated competitive binding inhibition and flow cytometry indicated a good specificity and sensitivity of **71** toward hypoxic cell detection. The probe showed low toxicity and, more importantly, the in vivo results showed a rapid response in tumor recognition and monitoring of liver cancer, enteritis, and liver ischemia.

Probe **72** is a BODIPY-based fluorogenic compound with quenched emissive properties due to the presence of a 4-nitrobenzyl fragment. The selective reduction of **72** in the presence of NTR results in the formation of a fluorescent compound following the 4-nitrobenzyl scavenging reaction. A 20-fold increase in fluorescent emission was observed after the addition of 1 μg/mL NTR. The detection limit of **72**, calculated according to regression analysis, was found to be 1.52 ng/mL NTR. The probe was nontoxic, with $\geq$90% cell viability after incubation with 0–50 μM for 24 h. The in vivo NIR optical imaging of CT26 solid tumor-bearing mice suggests that **72** could serve as a tumor-targeting, hypoxia-activatable probe for direct cancer monitoring both in vitro and in vivo.

In recent times, there has been significant attention given to two-photon fluorescent probes due to their advantages over traditional one-photon probes, including deeper tissue imaging depth, higher spatial resolution, and longer observation times. This motivated Zhai et al. and Wang et al. to synthesize two-photon probes **73** (Figure 33) and **75**

(Figure 34), which incorporate the 4-nitrobenzylic receptor fragment for selective recognition of nitroreductase (NTR) [78,79]. Compound **73** exhibited a remarkable 130-fold fluorescence enhancement at 563 nm within 10 min of reduction by NTR, with a detection limit of 23.67 ng/mL. It was successfully employed for imaging NTR activity in living HeLa cells, tissues, and zebrafish under hypoxic conditions. Notably, in a rat liver tumor model, probe **73** produced bright fluorescence even at a tissue depth of 200 μm, highlighting its effectiveness in deep-tissue imaging scenarios.

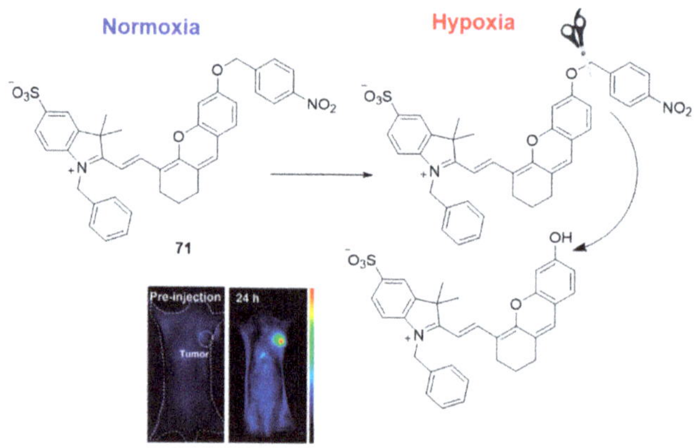

Figure 31. Fluorescence sensing mechanism of probe **71** for detection of hypoxia, and in vivo hypoxia-activated tumor imagining with **71**. Adapted with permission from [76]. Copyright (2022) Elsevier.

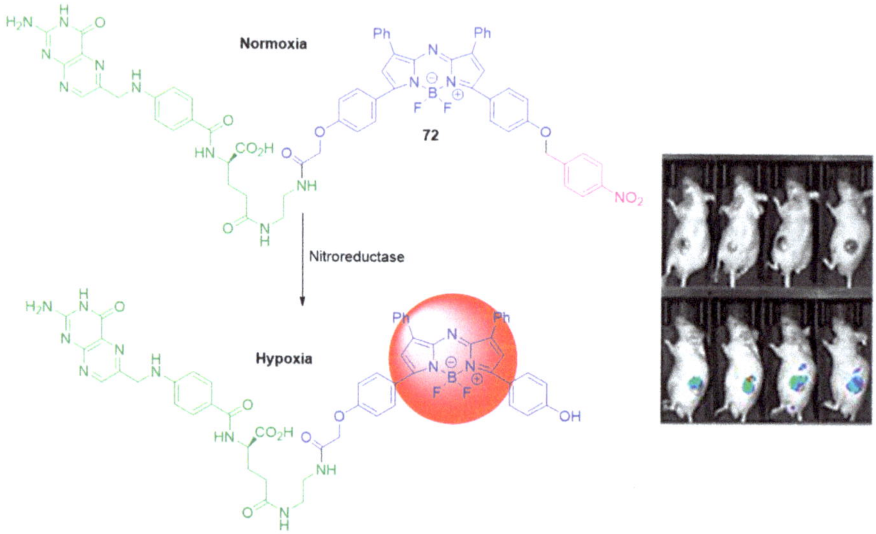

Figure 32. Fluorescence sensing mechanism of probe **72** for detection of hypoxia, and in vivo hypoxia-activated tumor imagining with **72**. Adapted with permission from [77]. Copyright (2021) American Chemical Society.

Similarly to **73**, probe **75** exhibits significant fluorescence enhancement upon selective reduction by NTR, albeit at a wavelength centered around 580 nm. The quantum yield of fluorescence post-reduction was notably increased to 0.045, compared to only 0.001 in its initial state. There exists a strong linear correlation between the fluorescence enhancement and NTR concentrations within the range of 0–20 μg/mL, with a calculated limit of detection of 26 ng/mL. Probe **75** has been effectively utilized for hypoxia imaging in A549 cell lines and A549 xenograft mice models, demonstrating its practical application in biological settings. Furthermore, MTT assays have revealed low toxicity associated with probe **75**, supporting its potential for safe use in biological and medical research contexts.

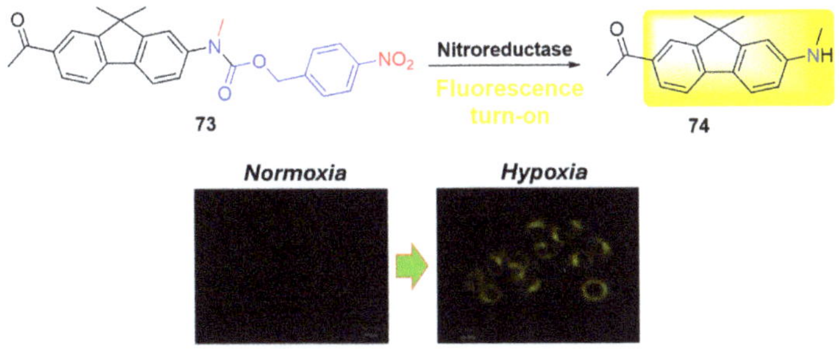

Figure 33. Fluorescence sensing mechanism of probe **73** for detection of NTR and HepG2 cells incubated with **73** under normoxic and hypoxic conditions. Adapted with permission from [78]. Copyright (2017) The Royal Society of Chemistry.

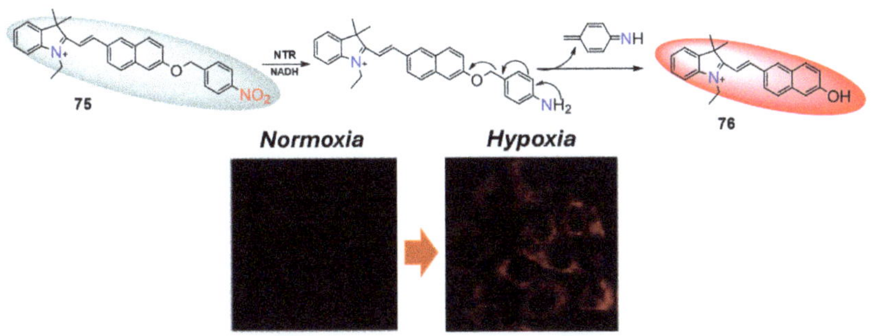

Figure 34. Fluorescence sensing mechanism of probe **75** for detection of NTR and A549 cells incubated with **75** under normoxic and hypoxic conditions. Adapted with permission from [79]. Copyright (2020) Elsevier.

Some nitro heterocycle side chains, similar to the 4-nitrobenzylic and 4-nitrobenzyl formiatic units discussed earlier, interact with NTR through scavenger reduction, making them promising recognition components in the design of fluorescent probes for detecting and imaging NTR. For example, phenoxazinone **77**, depicted in Figure 35, contains a 5-nitrofuranyl moiety that undergoes scavenger reduction, resulting in the formation of a highly fluorescent compound after the furane group is removed [80]. This reduction process is selective to NTR and enables its fluorescent detection. Probe **77** exhibits a remarkable 100-fold fluorescence enhancement and demonstrates a limit of detection of 0.27 ng/mL. Its potential for tumor diagnosis via hypoxia imaging was demonstrated by monitoring the hypoxic status in HeLa and A549 cells. Furthermore, standard MTT assays revealed that cell viability remained unaffected even at high concentrations (up to 5 μM) of **77**, indicating its low toxicity profile.

Probe **78**, as reported by Feng et al. (Figure 36), is another interesting example in which a 4-nitroimidazole recognition unit could be selectively removed after reduction with nitroreductase [81]. Probe **78** itself is nonfluorescent due to the photoinduced electron transfer to the nitroimidazole. However, after reduction by NTR, it was converted to 4-hydroxy-3-hydroxyflavone, which exhibits bright fluorescence at 560 nm due to ESIPT (intramolecular proton transfer). The ESIPT process showed an unusually high Stokes shift, which is a serious advantage in fluorescence sensing measurements, especially in living systems, because it could reduce the influence of unwanted self-reabsorption and the inner-filter effect. Due to the higher sensitivity and selective turn-on fluorescence response, probe **78** showed high potential for the detection of NTR, with good linearity in the concentration range of NTR 1–4 μg/mL and a limit of detection of 63 ng/mL. In addition, this probe displayed low cytotoxicity, good biocompatibility, and was successfully applied for imaging the hypoxic status of HeLa cells.

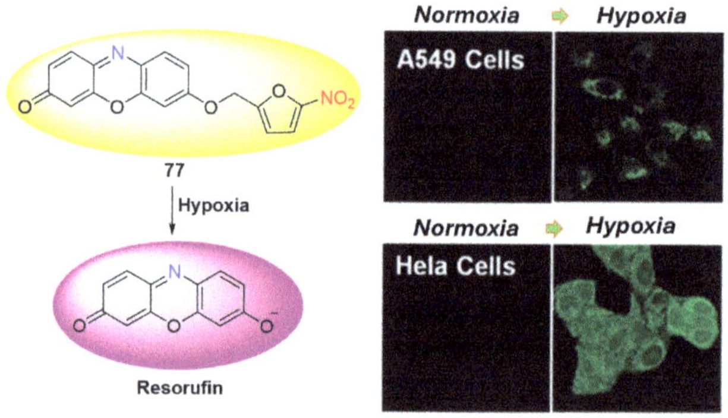

Figure 35. Fluorescence sensing mechanism of probe **77** for detection of NTR and Hela and A549 cells incubated with **77** under normoxic and hypoxic conditions. Adapted with permission from [80]. Copyright (2013) American Chemical Society.

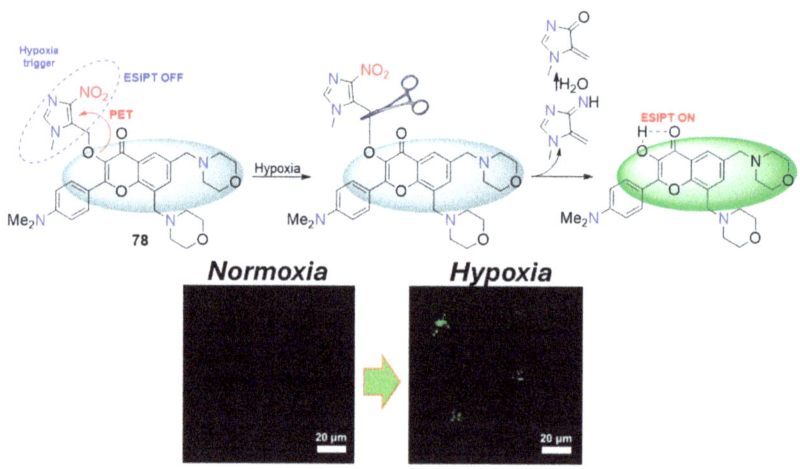

Figure 36. Fluorescence sensing mechanism of probe **78** for detection of NTR and Hela cells incubated with **78** under normoxic and hypoxic conditions. Adapted with permission from [81]. Copyright (2016) Elsevier.

In the last decade, the design and synthesis of selectively activated fluorescent probes in the second near-infrared window (NIR-II) has focused in general on the in vivo imagining of various biological or pathological processes, since NIR-II fluorescence imaging has an improved penetration depth and reduced autofluorescence. This motivated Meng et. al. to fabricate molecular probe **79** (Figure 37), which exhibits weak fluorescence at the NIR-II region due to the presence of a nitroimidazole fluorescent quencher [82]. The selective reduction of the nitro group to an amine in the presence of NTR resulted in a 107-fold fluorescent enhancement at 1046 nm. The NIR-II fluorescence signal of **78** in the tumor tissue was clearly visible at 10 h post-injection and reached its maximum at about 14 h. Notably, the in vivo

NIR-II fluorescence imaging of tumor hypoxia with 78 showed an unusual lack of observable background signal. The probe itself was safe, and it was found that it could be eliminated by renal excretion pathways from the animal body. All observed results revealed the great importance of probe **79** as a promising contrast and theranostic agent for hypoxia-related diseases, such as cancer, inflammation, stroke, and cardiac ischemia.

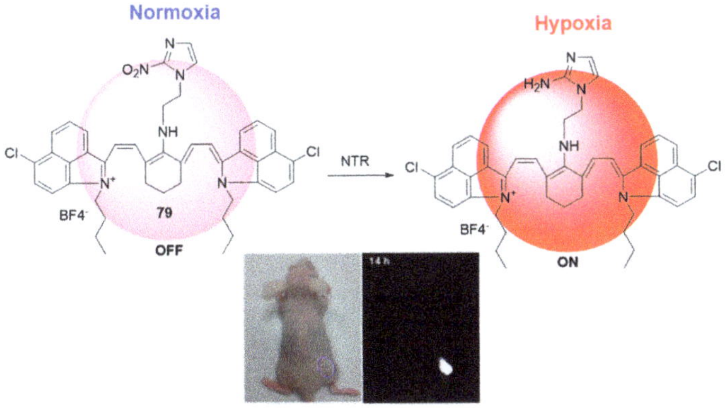

Figure 37. Fluorescence sensing mechanism of probe **79** for detection of hypoxia, and in vivo hypoxia-activated tumor imagining with **79**. Adapted with permission from [82]. Copyright (2018) Ivyspring International Publisher.

Karan et. al. reported another NIR II fluorescent probe (**80**) using 4-nitrobenzyl fragment as the NTR recognition element [83]. This probe was characterized by selective and a 4-fold more intensive ratiometric (1000 nm/940 nm) fluorescent output after activation by scavenging reduction of the 4-nitrobenzyl moiety. The ratiometric fluorescent response takes 45 min to reach its maximal value. It shows an excellent linear relationship against the concentrations of NTR from 0 to 10 μg/mL. The in vivo study indicated that probe **80** could be used for the visualization of tumor tissues due to the selective activation by NTR. Female BALB/c mice bearing two subcutaneous 4T1 breast tumors were chosen as a model. This probe offers a valuable in vivo biosensing amalgamation of NIR-II fluorescent response and self-calibrated ratiometric analysis (see Figure 38).

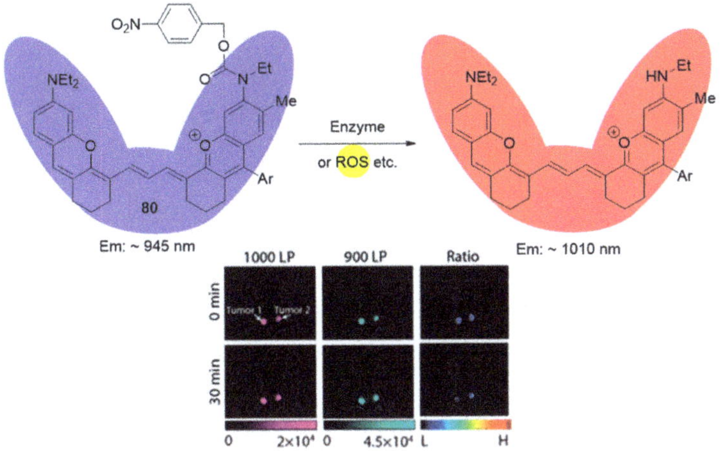

Figure 38. Fluorescence sensing mechanism of probe **80** for detection of hypoxia, and in vivo hypoxia-activated tumor imagining with **80**. Adapted with permission from [83]. Copyright (2022) American Chemical Society.

5. Conclusions

The heterogeneity of tumor hypoxia, pharmacokinetics, and the potential toxicity of conventional drugs remain significant challenges for HAPs. Continued research into nitro(het)aromatic compounds and their structural modifications is essential for developing more selective and effective cancer therapies. This review underscores the potential of hypoxia-activated nitro(het)aromatic prodrugs in targeted cancer therapy, particularly through mechanisms such as gene-directed enzyme prodrug therapy and the formulation of bioreductive-activated prodrug conjugates featuring nitrophenyl, nitrobenzyl, and nitro-heteroaryl triggers. While promising, some BAPCs require further refinement to reduce activation in normoxic conditions and improve water solubility. To address these challenges, the design of next-generation hypoxia-activated pro-drugs focuses on incorporating bulkier nitroaromatic groups and triggers with higher reduction potentials, which better match the cellular reductase range. This approach aims to improve specificity and minimize the off-target effects, thereby maximizing drug delivery and activation specifically within hypoxic tumor regions. Overall, BAPCs represent a significant advancement in targeted cancer therapy, offering a method to specifically deliver and activate drugs in hypoxic tumor regions, thereby enhancing efficacy and minimizing systemic toxicity. Stability and bioavailability issues arise from poor compound stability in biological environments and limited bioavailability. To address these, chemical modifications to enhance stability and formulation strategies, including encapsulation in liposomes or other nanocarriers, are recommended.

Over the past four decades, nitroaromatic compounds have played a crucial role in the design and synthesis of fluorescent probes for the selective detection of NTR and for imaging the hypoxic status of tumor cells and tissues. The major appeal of nitroaromatic derivatives lies in their simple synthesis and predictable off-on fluorescent response after selective bioreduction by NTR. Many of the developed probes exhibit low toxicity, high sensitivity, and linearity in detection ranges, making them suitable for in vivo applications in the hypoxia imaging of tumors. Despite significant progress, enhancing water solubility remains a critical hurdle due to the inherently hydrophobic nature of most organic architectures, limiting their practical application.

Future research directions include developing multianalyte probes capable of detecting acidity and NTR simultaneously, thereby reducing false positives in clinical settings. Moreover, exploring two-photon fluorescent technologies for deeper tissue imaging and improved diagnostic accuracy holds promise. These advancements highlight the potential of nitro(het)aromatic compounds in revolutionizing cancer diagnostics and therapy, offering reliable, non-invasive solutions for early cancer detection and treatment monitoring.

We hope this review will assist researchers in creating novel nitro(het)aromatic compounds for treating with improve selectivity and stability, and/or for fluorescently imaging hypoxic tumors to bolster cancer diagnostics and improve the monitoring of treatment efficacy.

Author Contributions: Conceptualization, K.A., D.Y. and N.G.; writing—original draft preparation, N.L., V.B. and D.Y.; writing—review and editing, K.A. and N.G.; supervision, N.G.; design, K.A.; visualization, V.B. All authors have read and agreed to the published version of the manuscript.

Funding: This study is funded by the European Union—NextGenerationEU, through the National Recovery and Resilience Plan of the Republic of Bulgaria, project № BG-RRP-2.004-0002, "BiOrgaMCT".

Institutional Review Board Statement: Not applicable.

Data Availability Statement: Data are available within the article.

Conflicts of Interest: The authors declare no conflicts of interest.

References

1. Muz, B.; de la Puente, P.; Azab, F.; Azab, A.K. The role of hypoxia in cancer progression, angiogenesis, metastasis, and resistance to therapy. *Hypoxia* **2015**, *3*, 83–92. [CrossRef]
2. Hompland, T.; Fjeldbo, C.S.; Lyng, H. Tumor Hypoxia as a Barrier in Cancer Therapy: Why Levels Matter. *Cancers* **2021**, *13*, 499. [CrossRef]
3. Chen, Z.; Han, F.; Du, Y.; Shi, H.; Zhou, W. Hypoxic microenvironment in cancer: Molecular mechanisms and therapeutic interventions. *Signal Transduct. Target. Ther.* **2023**, *8*, 70. [CrossRef]
4. Jing, X.; Yang, F.; Shao, C.; Wei, K.; Xie, M.; Shen, H.; Shu, Y. Role of hypoxia in cancer therapy by regulating the tumor microenvironment. *Mol. Cancer* **2019**, *18*, 157. [CrossRef]
5. Rakotomalala, A.; Escande, A.; Furlan, A.; Meignan, S.; Lartigau, E. Hypoxia in Solid Tumors: How Low Oxygenation Impacts the "Six Rs" of Radiotherapy. *Front. Endocrinol.* **2021**, *12*, 742215. [CrossRef]
6. Denny, W.A.; Wilson, W.R.; Hay, M.P. Recent developments in the design of bioreductive drugs. *Br. J. Cancer Suppl.* **1996**, *74*, S32–S38.
7. Phillips, R.M. Targeting the hypoxic fraction of tumours using hypoxia-activated prodrugs. *Cancer Chemother. Pharmacol.* **2016**, *77*, 441–457. [CrossRef]
8. Lin, A.J.; Pardini, R.S.; Cosby, L.A.; Lillis, B.J.; Shansky, C.W.; Sartorelli, A.C. Potential bioreductive alkylating agents. 2. Antitumor effect and biochemical studies of naphthoquinone derivatives. *J. Med. Chem.* **1973**, *16*, 1268–1271. [CrossRef]
9. Guise, C.P.; Mowday, A.M.; Ashoorzadeh, A.; Yuan, R.; Lin, W.-H.; Wu, D.-H.; Jeff, B.; Smaill, J.B.; Adam, V.; Patterson, A.; et al. Bioreductive prodrugs as cancer therapeutics: Targeting tumor hypoxia. *Chin. J. Cancer* **2014**, *33*, 80–86. [CrossRef]
10. Nepali, K.; Lee, H.-Y.; Liou, J.-P. Nitro Group Containing Drugs. *J. Med. Chem.* **2019**, *62*, 1701–1714. [CrossRef]
11. Spiegelberg, L.; Houben, R.; Niemans, R.; de Ruysscher, D.; Yaromina, A.; Theys, J.; Guise, C.P.; Smaill, J.B.; Patterson, A.V.; Lambin, P.; et al. Hypoxia-activated prodrugs and (lack of) clinical progress: The need for hypoxia-based biomarker patient selection in phase III clinical trials. *Clin. Transl. Radiat. Oncol.* **2019**, *15*, 62–69. [CrossRef]
12. Weiss, G.J.; Infante, J.R.; Chiorean, E.G.; Borad, M.J.; Bendell, J.C.; Molina, J.R.; Tibes, R.; Ramanathan, R.K.; Lewandowski, K.; Jones, S.F.; et al. Phase 1 study of the safety, tolerability, and pharmacokinetics of TH-302, a hypoxia-activated prodrug, in patients with advanced solid malignancies. *Clin. Cancer Res.* **2011**, *17*, 2997–3004. [CrossRef]
13. Borad, M.J.; Reddy, S.G.; Bahary, N.; Uronis, H.E.; Sigal, D.; Cohn, A.L.; Schelman, W.R.; Jr, J.S.; Chiorean, E.G.; Rosen, P.J.; et al. Randomized phase II trial of gemcitabine plus TH-302 versus gemcitabine in patients with advanced pancreatic cancer. *J. Clin. Oncol.* **2015**, *33*, 1475–1481. [CrossRef]
14. Tap, W.D.; Papai, Z.; Van Tine, B.A.; Attia, S.; Ganjoo, K.N.; Jones, R.L.; Schuetze, S.; Reed, D.; Chawla, S.P.; Riedel, R.F.; et al. Doxorubicin plus evofosfamide versus doxorubicin alone in locally advanced, unresectable or metastatic soft-tissue sarcoma (TH CR-406/SARC021): An international, multicentre, open-label, randomised phase 3 trial. *Lancet Oncol.* **2017**, *18*, 1089–1103. [CrossRef]
15. Konopleva, M.; Thall, P.F.; Yi, C.A.; Borthakur, G.; Coveler, A.; Bueso-Ramos, C.; Benito, J.; Konoplev, S.; Gu, Y.; Ravandi, F.; et al. Phase I/II study of the hypoxia-activated prodrug PR104 in refractory/relapsed acute myeloid leukemia and acute lymphoblastic leukemia. *Haematologica* **2015**, *100*, 927–934. [CrossRef]
16. McKeage, M.J.; Jameson, M.B.; Ramanathan, R.K.; Rajendran, J.; Gu, Y.; Wilson, W.R.; Melink, T.J.; Tchekmedyian, N.S. PR-104 a bioreductive pre-prodrug combined with gemcitabine or docetaxel in a phase Ib study of patients with advanced solid tumours. *BMC Cancer* **2012**, *12*, 496. [CrossRef]
17. Denny, W.A. Nitroaromatic Hypoxia-Activated Prodrugs for Cancer Therapy. *Pharmaceuticals* **2022**, *15*, 187. [CrossRef]
18. Macherey, A.; Dansette, P. Biotransformations Leading to Toxic Metabolites: Chemical Aspects. In *The Practice of Medicinal Chemistry*, 4th ed.; Wermuth, C.G., Aldous, D., Raboisson, P., Rognan, D., Eds.; Academic Press of Elsevier: London, UK, 2015; pp. 585–614.
19. Patterson, A.V.; Ferry, D.M.; Edmunds, S.J.; Gu, Y.; Singleton, R.S.; Patel, K.; Pullen, S.M.; Hicks, K.O.; Syddall, S.P.; Atwell, G.J.; et al. Mechanism of Action and Preclinical Antitumor Activity of the Novel Hypoxia-Activated DNA Cross-Linking Agent PR-104. *Clin. Cancer Res.* **2007**, *13*, 3922–3932. [CrossRef]
20. Shah, K.; Chauhan, D.N.; Chauhan, N.S.; Mishra, P. *Recent Advancement in Prodrugs*, 1st ed.; CRC Press: Boca Raton, FL, USA, 2020; pp. 227–252.
21. McKeown, S.R.; Coweny, R.L.; Williamsy, K.J. Bioreductive Drugs: From Concept to Clinic. *Clin. Oncol.* **2007**, *19*, 427–442. [CrossRef]
22. Denny, W.A. Hypoxia-activated prodrugs in cancer therapy: Progress to the clinic. *Future Oncol.* **2010**, *6*, 419–428. [CrossRef]
23. Movsas, B.; Chapman, J.D.; Hanlon, A.L. Hypoxic prostate/muscle po_2 ratio predicts for biochemical failure in patients with prostate cancer: Preliminary findings. *Urology* **2002**, *60*, 634–639. [CrossRef]
24. Zhang, J.; Kale, V.; Chen, M. Gene-Directed Enzyme Prodrug Therapy. *AAPS J.* **2014**, *17*, 102–110. [CrossRef]
25. Çelik, A.; Yetiş, G. An unusually cold active nitroreductase for prodrug activations. *Bioorg. Med. Chem.* **2012**, *20*, 3540–3550. [CrossRef] [PubMed]
26. Güngör, T.; Yetis, G.; Onder, F.C.; Tokay, E.; Tok, T.; Celik, A.; Ay, M.; Kockar, F. Prodrugs For Nitroreductase Based Cancer Therapy-1 Metabolite Profile, Cell Cytotoxicity and Molecular Modeling Interactions of Nitro Benzamides with Ssap-NtrB. *Med. Chem.* **2018**, *14*, 495–507. [CrossRef]

27. Güngör, T.; Önder, F.C.; Tokay, E.; Gülhan, Ü.G.; Hacıoğlu, N.; Tok, T.T.; Çelik, A.; Koçkar, F.; Ay, M. Prodrugs for nitroreductase based cancer therapy-2: Novel amide/Ntr combinations targeting PC3 cancer cells. *Eur. J. Med. Chem.* **2019**, *171*, 383–400. [CrossRef] [PubMed]
28. Tokay, E.; Güngör, T.; Hacıoğlu, N.; Önder, F.C.; Gülhan, Ü.G.; Tok, T.T.; Çelik, A.; Ay, M.; Köçkar, F. Prodrugs for nitroreductase-based cancer therapy-3: Antitumor activity of the novel dinitroaniline prodrugs/Ssap-NtrB enzyme suicide gene system: Synthesis, in vitro and in silico evaluation in prostate cancer. *Eur. J. Med. Chem.* **2019**, *187*, 111937. [CrossRef]
29. Güngör, T.; Tokay, E.; Güven Gülhan, Ü.; Hacıoğlu, N.; Çelik, A.; Köçkar, F.; Ay, M. Prodrugs for nitroreductase based cancer therapy-4: Towards prostate cancer targeting: Synthesis of N-heterocyclic nitro prodrugs, Ssap-NtrB enzymatic activation and anticancer evaluation. *Bioorg. Chem.* **2020**, *105*, 104450. [CrossRef] [PubMed]
30. Hacıoğlu, N.; Güngör, T.; Tokay, E.; Gülhan, Ü.G.; Çelik, A.; Ay, M.; Köçkar, F. Prodrugs for Nitroreductase Based Cancer Therapy-5: Development of Trinitroaniline Prodrugs/Ssap-NtrB Combinations for Liver Cancer Using Intracellular and Extracellular Conditions. *ChemistrySelect* **2021**, *6*, 6315–6323. [CrossRef]
31. Luo, X.; Li, A.; Chi, X.; Lin, Y.; Liu, X.; Zhang, L.; Su, X.; Yin, Z.; Yin, Z.; Lin, H.; et al. Hypoxia-Activated Prodrug Enabling Synchronous Chemotherapy and HIF-1α Downregulation for Tumor Treatment. *Bioconjugate Chem.* **2021**, *32*, 983–990. [CrossRef]
32. Kang, D.; Cheung, S.T.; Wong-Rolle, A.; Kim, J. Enamine N-Oxides: Synthesis and Application to Hypoxia-Responsive Prodrugs and Imaging Agents. *ACS Cent. Sci.* **2021**, *7*, 631–640. [CrossRef]
33. Anduran, E.; Dubois, L.J.; Lambin, P.; Winum, J.Y. Hypoxia-Activated Prodrug Derivatives of Anti-Cancer Drugs: A Patent Review 2006–2021. *Expert Opin. Ther. Pat.* **2022**, *32*, 1–12. [CrossRef] [PubMed]
34. Meng, F.; Bhupathi, D.; Sun, J.D.; Liu, Q.; Ahluwalia, D.; Wang, Y.; Matteucci, M.D.; Hart, C.P. Enhancement of hypoxia-activated prodrug TH-302 anti-tumor activity by Chk1 inhibition. *BMC Cancer* **2015**, *15*, 422. [CrossRef] [PubMed]
35. Singh, Y.; Palombo, M.; Sinko, P. Recent Trends in Targeted Anticancer Prodrug and Conjugate Design. *Curr. Med. Chem.* **2008**, *15*, 1802–1826. [CrossRef] [PubMed]
36. Damen, E.W.P.; Nevalainen, T.J.; Van den Bergh, T.J.; de Groot, F.M.H.; Scheeren, H.W. Synthesis of novel paclitaxel prodrugs designed for bioreductive activation in hypoxic tumour tissue. *Bioorg. Med. Chem.* **2002**, *10*, 71–77. [CrossRef]
37. Ikeda, Y.; Hisano, H.; Nishikawa, Y.; Nagasaki, Y. Targeting and Treatment of Tumor Hypoxia by Newly Designed Prodrug Possessing High Permeability in Solid Tumors. *Mol. Pharm.* **2016**, *13*, 2283–2289. [CrossRef] [PubMed]
38. Mahato, R.; Tai, W.; Cheng, K. Prodrugs for improving tumor targetability and efficiency. *Adv. Drug Deliv. Rev.* **2011**, *63*, 659–670. [CrossRef]
39. Al-Hilal, T.A.; Hossain, M.A.; Alobaida, A.; Alam, F.; Keshavarz, A.; Nozik-Grayck, E.; Stenmark, K.R.; German, N.A.; Ahsan, F.J. Design, synthesis and biological evaluations of a long-acting, hypoxia-activated prodrug of fasudil, a ROCK inhibitor, to reduce its systemic side-effects. *J. Control. Release* **2021**, *334*, 237–247. [CrossRef]
40. Battello, N.; Zimmer, A.D.; Goebel, C.; Dong, X.; Behrmann, I.; Haan, C.; Hiller, K.; Wegner, A. The role of HIF-1 in oncostatin M-dependent metabolic reprogramming of hepatic cells. *Cancer Metab.* **2016**, *4*, 3. [CrossRef] [PubMed]
41. Shin, D.H.; Kim, J.H.; Jung, Y.J.; Kim, K.E.; Jeong, J.M.; Chun, Y.S.; Park, J.W. Preclinical evaluation of YC-1, a HIF inhibitor, for the prevention of tumor spreading. *Cancer Lett.* **2007**, *255*, 107–116. [CrossRef]
42. Zhang, B.; Huang, X.; Wang, H.; Gou, S. Promoting antitumor efficacy by suppressing hypoxia via nano self-assembly of two irinotecan-based dual drug conjugates having a HIF-1alpha inhibitor. *J. Mater. Chem.* **2019**, *B7*, 5352–5362. [CrossRef]
43. Ge, Y.; Lai, X.; Li, J.; Yu, R.; Zhuang, Z.; Sun, G.; Cui, X.; Zhang, N.; Zhao, L.; Upadhyaya, P.; et al. NBGNU: A hypoxia-activated tripartite combinitrosourea prodrug overcoming AGT-mediated chemoresistance. *Future Med. Chem.* **2019**, *11*, 269–284. [CrossRef] [PubMed]
44. Liang, D.; Wu, X.; Hasinoff, B.B.; Herbert, D.E.; Tranmer, G.K. Evaluation of Ni-trobenzyl Derivatives of Camptothecin as Anti-Cancer Agents and Potential Hypoxia Targeting Prodrugs. *Molecules* **2018**, *23*, 2041. [CrossRef] [PubMed]
45. Jin, C.; Zhang, Q.; Lu, W. Synthesis and biological evaluation of hypoxia-activated prodrugs of SN-38. *Eur. J. Med. Chem.* **2017**, *132*, 135–141. [CrossRef] [PubMed]
46. Bielec, B.; Schueffl, H.; Terenzi, A.; Berger, W.; Heffeter, P.; Keppler, B.K.; Kowol, C.R. Development and biological investigations of hypoxia-sensitive prodrugs of the tyrosine kinase inhibitor crizotinib. *Bioorg. Chem.* **2020**, *99*, 103778. [CrossRef]
47. Huang, Y.; Jin, C.; Yu, J.; Wang, L.; Lu, W. A Novel Multifunctional 2-Nitroimidazole-based Bioreductive Linker and its Application in Hypoxia-Activated Prodrugs. *Bioorg. Chem.* **2020**, 103975. [CrossRef]
48. Cheng, Z.; Huang, Y.; Shao, P.; Wang, L.; Zhy, S.; Yu, J.; Wei, L. Hypoxia-Activated Albumin-Binding Exatecan Prodrug for Cancer Therapy. *ACS Omega* **2022**, *7*, 1082–1089. [CrossRef] [PubMed]
49. Zhang, Q.; Jin, C.; Yu, J.; Lu, W. Synthesis of New Branched 2-Nitroimidazole as a Hypoxia Sensitive Linker for Ligand-Targeted Drugs of Paclitaxel. *ACS Omega* **2018**, *3*, 8813–8818. [CrossRef]
50. Winn, B.A.; Devkota, L.; Kuch, B.; MacDonough, M.T.; Strecker, T.E.; Wang, Y.; Pinney, K.G. Bioreductively Activatable Prodrug Conjugates of Combretastatin A-1 and Combretastatin A-4 as Anticancer Agents Targeted toward Tumor-Associated Hypoxia. *J. Nat. Prod.* **2020**, *83*, 937–954. [CrossRef] [PubMed]
51. Kumari, R.; Sunil, D.; Ningthoujam, R.S. Naphthalimides in fluorescent imaging of tumor hypoxia—An up-to-date Review. *Bioorg. Chem.* **2019**, *88*, 102979. [CrossRef]
52. Elmes, R.B.P. Bioreductive fluorescent imaging agents: Applications to tumour hypoxia. *Chem. Commun.* **2016**, *52*, 8935–8956. [CrossRef]

53. Janczy-Cempa, E.; Mazuryk, O.; Kania, A.; Brindell, M. Significance of Specific Oxidoreductases in the Design of Hypoxia-Activated Prodrugs and Fluorescent Turn off–on Probes for Hypoxia Imaging. *Cancers* **2022**, *14*, 2686. [CrossRef] [PubMed]
54. Mirabello, V.; Cortezon-Tamarit, F.; Pascu, S.I. Oxygen Sensing, Hypoxia Tracing and in Vivo Imaging with Functional Metalloprobes for the Early Detection of Non-communicable Diseases. *Front. Chem.* **2018**, *6*, 27.
55. Qiao, J.; Wang, M.; Cui, M.; Fang, Y.; Li, H.; Zheng, C.; Li, Z.; Xu, Y.; Hua, H.; Li, D. Small-molecule probes for fluorescent detection of cellular hypoxia-related nitroreductase. *J. Pharm. Biomed. Anal.* **2021**, *203*, 114199. [CrossRef]
56. Qi, Y.-L.; Guo, L.; Chen, L.-L.; Li, H.; Yang, Y.-S.; Jiang, A.-Q.; Zhu, H.-L. Recent progress in the design principles, sensing mechanisms, and applications of small-molecule probes for nitroreductases. *Coord. Chem. Rev.* **2020**, *421*, 213460. [CrossRef]
57. Godet, I.; Doctorman, S.; Wu, F.; Gilkes, D. Detection of Hypoxia in Cancer Models: Significance, Challenges, and Advances. *Cells* **2022**, *11*, 686. [CrossRef]
58. Zhang, X.; Li, S.; Ma, H.; Wang, H.; Zhang, R.; Zhang, X.-D. Activatable NIR-II organ-ic fluorescent probes for bioimaging. *Theranostics* **2022**, *12*, 3345. [CrossRef]
59. Liu, J.; Bu, W.; Shi, J. Chemical Design and Synthesis of Functionalized Probes for Imaging and Treating Tumor Hypoxia. *Chem. Rev.* **2017**, *117*, 6160–6224. [CrossRef]
60. Aguilera, K.; Brekken, R. Hypoxia Studies with Pimonidazole in vivo. *Bio Protoc.* **2014**, *4*, e1254. [CrossRef]
61. Olive, P.L.; Durand, R.E. Fluorescent Nitroheterocycles for identifying hypoxic cells. *Cancer Res.* **1983**, *43*, 3276–3280.
62. Olive, P.L. Cellular metabolism of fluorescent nitroheterocycles. *Int. J. Radiat. Oncol. Biol. Phys.* **1984**, *10*, 1357–1360. [CrossRef]
63. Liu, Y.; Xu, Y.; Qian, X.; Liu, J.; Shen, L.; Lic, J.; Zhang, Y. Novel fluorescent markers for hypoxic cells of naphthalimides with two heterocyclic side chains for bioreductive binding. *Bioorg. Med. Chem.* **2006**, *14*, 2935–2941. [CrossRef] [PubMed]
64. Zhu, W.; Dai, M.; Xu, Y.; Qian, X. Novel nitroheterocyclic hypoxic markers for solid tumor: Synthesis and biological evaluation. *Bioorg. Med. Chem.* **2008**, *16*, 3255–3260. [CrossRef] [PubMed]
65. Dai, M.; Zhu, W.; Xu, Y.; Qian, X.; Liu, Y.; Xiao, Y.; You, Y. Versatile Nitro-Fluorophore as Highly Effective Sensor for Hypoxic Tumor Cells: Design, Imaging and Evaluation. *J. Fluoresc.* **2008**, *18*, 591–597. [CrossRef] [PubMed]
66. Fan, L.; Zan, Q.; Lin, B.; Wang, X.; Gong, X.; Zhao, Z.; Shuang, S.; Dong, C.; Wong, M. Hypoxia imaging in living cells, tissues and zebrafish with a nitroreductase-specific fluorescent probe. *Analyst* **2020**, *145*, 5657–5663. [CrossRef] [PubMed]
67. Janczy-Cempa, E.; Mazuryk, O.; Sirbu, D.; Chopin, N.; Żarnik, M.; Zastawna, M.; Colas, C.; Hiebel, M.-A.; Suzenet, F.; Brindell, M. Nitro-Pyrazinotriazapentalene scaffolds–nitroreductase quantification and in vitro fluorescence imaging of hypoxia. *Sens. Actuators B Chem.* **2021**, *346*, 130504. [CrossRef]
68. Fan, Y.; Lu, M.; Yu, X.; He, M.; Zhang, Y.; Ma, X.; Kou, J.; Yu, B.; Tian, J. Targeted Myocardial Hypoxia Imaging Using a NitroreductaseActivatable Near-Infrared Fluorescent Nanoprobe. *Anal. Chem.* **2019**, *91*, 6585–6592. [CrossRef] [PubMed]
69. Zhang, Y.; Zhang, X.-F.; Chen, Q.; Cao, X.-Q.; Shen, S.-L. A novel near-infrared fluorescence off-on probe for imaging hypoxia and nitroreductase in cells and in vivo A novel near-infrared fluorescence off-on probe for imaging hypoxia and nitroreductase in cells and in vivo. *Sens. Actuator* **2022**, *353*, 131145. [CrossRef]
70. Chen, S.; Yu, S.; Du, Z.; Huang, X.; He, M.; Long, S.; Liu, J.; Lan, Y.; Yang, D.; Wang, H.; et al. Synthesis of Mitochondria-Anchored Nitroimidazoles with a Versatile NIR Fluorophore for Hypoxic Tumor-Targeting Imaging and Chemoradiotherapy. *J. Med. Chem.* **2021**, *64*, 3381–3391. [CrossRef]
71. Zhou, Y.; Bobba, K.N.; Lv, X.W.; Yang, D.; Velusamy, N.; Zhang, J.F.; Bhuniya, S. A biotinylated piperazine-rhodol derivative: A "turn-on" probe for nitroreductase triggered hypoxia imaging. *Analyst* **2017**, *142*, 345–350. [CrossRef]
72. Wei, C.; Shen, Y.; Xu, Z.; Peng, S.; Yuan, Z.; He, Y.; Yin, J.; Chen, H. A novel off-on fluorescent probe for imaging of hypoxia in tumor cells. *J. Photochem. Photobiol. A* **2018**, *353*, 292–298. [CrossRef]
73. He, Z.; Chou, Y.; Zhou, H.; Zhang, H.; Cheng, T.; Liu, G. A nitroreductase and acidity detecting dual functional ratiometric fluorescent probe for selectively imaging tumor cells. *Org. Biomol. Chem.* **2018**, *16*, 3266–3272. [CrossRef] [PubMed]
74. Zheng, J.; Shen, Y.; Xu, Z.; Yuan, Z.; He, Y.; Wei, C.; Er, M.; Yin, J.; Chen, H. Near-infrared off-on fluorescence probe activated by NTR for in vivo hypoxia imaging. *Biosens. Bioelectron.* **2018**, *119*, 141–148. [CrossRef]
75. Xia, L.; Hu, F.; Huang, J.; Li, N.; Gu, Y.; Wang, P. A fluorescent turn-on probe for nitroreductase imaging in living cells and tissues under hypoxia conditions. *Sens. Actuators B Chem.* **2018**, *268*, 70–76. [CrossRef]
76. Lin, Q.; Li, C.; Wang, L.; Cai, H.; Tang, L.; Gu, Y. Ultrasensitive near-infrared fluorescence probe activated by nitroreductase for in vivo hypoxia detection. *Sens. Actuators B Chem.* **2022**, *371*, 132521. [CrossRef]
77. Karan, S.; Cho, M.; Lee, H.; Park, H.; Sundararajan, M.; Sessler, J.; Hong, K. Near-Infrared Fluorescent Probe Activated by Nitroreductase for In Vitro and In Vivo Hypoxic Tumor Detection. *J. Med. Chem.* **2021**, *64*, 2971–2981. [CrossRef]
78. Zhai, B.; Hu, W.; Sun, J.; Chi, S.; Lei, Y.; Zhang, F.; Zhong, C.; Liu, Z. A two-photon fluorescent probe for nitroreductase imaging in living cells, tissues and zebrafish under hypoxia conditions. *Analyst* **2017**, *142*, 1545–1553. [CrossRef] [PubMed]
79. Wang, Y.; Zhang, L.; Huang, Y.; Wang, X.; Zhang, L.; Chen, L. Rational design of a nitroreductase-activatable two-photon fluorescent probe for hypoxia imaging in cell and in vivo. *Sens. Actuators B Chem.* **2020**, *310*, 127755. [CrossRef]
80. Li, Z.; Li, X.; Gao, X.; Zhang, Y.; Shi, W.; Ma, H. Nitroreductase Detection and Hypoxic Tumor Cell Imaging by a Designed Sensitive and Selective Fluorescent Probe, 7-[(5-Nitrofuran-2-yl)methoxy]-3H-phenoxazin-3-one. *Anal. Chem.* **2013**, *85*, 3926–3932. [CrossRef]
81. Feng, W.; Wang, Y.; Chen, S.; Wang, C.; Wang, S.; Li, S.; Li, H.; Zhou, G.; Zhang, J. 4-Nitroimidazole-3-hydroxyflavone conjugate as a fluorescent probe for hypoxic cells. *Dyes Pigm.* **2016**, *131*, 145–153. [CrossRef]

82. Meng, X.; Zhang, J.; Sun, Z.; Zhou, L.; Deng, G.; Li, S.; Li, W.; Gong, P.; Cai, L. Hypoxia-triggered single molecule probe for high-contrast NIR II/PA tumor imaging and robust photothermal therapy. *Theranostics* **2018**, *8*, 6025–6034. [CrossRef]
83. Lan, Q.; Yu, P.; Yan, K.; Li, X.; Zhang, F.; Le, Z. Polymethine Molecular Platform for Ratiometric Fluorescent Probes in the Second near-Infrared Window. *J. Am. Chem. Soc.* **2022**, *144*, 21010–21015. [CrossRef] [PubMed]

molecules

MDPI

Review

Targeting Moonlighting Enzymes in Cancer

Chunxu Lin [1], Mingyang Yu [1], Ximei Wu [1], Hui Wang [1], Min Wei [1,*] and Luyong Zhang [1,2,*]

1 Center for Drug Research and Development, Guangdong Pharmaceutical University, Guangzhou 510006, China; 2112140131@gdpu.edu.cn (C.L.); 2112040080@gdpu.edu.cn (M.Y.); 2112240081@gdpu.edu.cn (X.W.); 2112240063@gdpu.edu.cn (H.W.)

2 Jiangsu Key Laboratory of Drug Screening, China Pharmaceutical University, Nanjing 210009, China

* Correspondence: min.wei@gdpu.edu.cn (M.W.); lyzhang@gdpu.edu.cn (L.Z.)

Abstract: Moonlighting enzymes are multifunctional proteins that perform multiple functions beyond their primary role as catalytic enzymes. Extensive research and clinical practice have demonstrated their pivotal roles in the development and progression of cancer, making them promising targets for drug development. This article delves into multiple notable moonlighting enzymes, including GSK-3, GAPDH, and ENO1, and with a particular emphasis on an enigmatic phosphatase, PTP4A3. We scrutinize their distinct roles in cancer and the mechanisms that dictate their ability to switch roles. Lastly, we discuss the potential of an innovative approach to develop drugs targeting these moonlighting enzymes: target protein degradation. This strategy holds promise for effectively tackling moonlighting enzymes in the context of cancer therapy.

Keywords: moonlighting enzyme; PTP4A3; targeted protein degradation; PROTAC

1. Introduction

Moonlighting proteins is the term used to describe a class of multifunctional proteins that perform more than one function within a cell, often without any change in their primary structure or sequence. The term "moonlighting" was coined by Dr. Constance Jeffery in 1999. In this seminal paper, Jeffery described moonlighting proteins as those that perform more than one function and do not have these functions as a consequence of gene fusions, families of homologous proteins, splice variants, or pleiotropic effects [1].

Since then, the comprehension of moonlighting proteins has undergone substantial development. The definition has been broadened to encompass proteins capable of transitioning between functions based on cellular conditions or executing distinct roles in various cellular compartments. A pivotal criterion for a protein to qualify as moonlighting has been elucidated: the autonomy of its diverse function, signifying that the disruption of one function (e.g., via mutation) should not influence the other function, and reciprocally. The notion of moonlighting proteins has gained widespread acceptance, as evidenced by a consistent influx of scholarly reviews focusing on this topic [2–10].

In the realm of enzymes, the term "moonlighting" pertains to proteins capable of executing multiple functions beyond their primary catalytic role. These proteins have evolved to encompass additional binding sites or functions while maintaining their original enzymatic activity. Enzymes stand out as compelling targets for drug development, with 50% of human protein drug targets belonging to the enzyme category [11]. They serve pivotal functions in diverse cellular processes and are intricately involved in disease pathways. Consequently, a significant number of current therapeutic approaches focus on inhibiting enzymes, with kinase and protease inhibitors being prominent examples that are widely utilized in clinical settings for the treatment of cancer.

Moonlighting enzymes challenge traditional views of protein specificity by showcasing how proteins can exhibit multiple functions beyond their primary roles. This versatility underscores the complexity of protein function and highlights the potential for developing novel therapeutic strategies targeting these moonlighting enzymes.

Moonlighting proteins are distributed across various species and can play crucial roles in disease development when their activities are altered. Moonlighting enzymes have been found to play critical roles in tumor growth and progression, making them potential targets for therapeutic intervention [6,9]. However, targeting moonlighting enzymes is challenging. Comprehensive understanding of the functions and regulations of moonlighting enzymes, and exploring innovative drug design strategies, are thus imperative for successful therapeutic intervention.

2. Moonlighting Enzymes Involved in Cancer

In this section, we aim to showcase specific instances of moonlighting enzymes that have been implicated in tumorigenesis. It is important to note that our selection represents only a subset of these enzymes, as previous reviews have provided comprehensive lists [6,9]. Additionally, we introduce a novel and emerging example, PTP4A3, which has not been extensively discussed in prior literature.

2.1. GSK-3

Glycogen synthase kinase 3 (GSK-3) is recognized as a serine/threonine kinase with a diverse range of substrates, including glycogen synthase, β-catenin, Cyclin D1, and c-Myc. These substrates are involved in important cellular processes, such as the cell cycle, apoptosis, and glycogen synthesis. The phosphorylation of glycogen synthase by GSK-3 inhibits its activity, thereby regulating glycogen metabolism [12,13]. Generally, the phosphorylation of GSK-3 substrates leads to the degradation of these proteins, which in turn results in the suppression of cell growth and proliferation [14,15].

Although GSK has been widely reported as a serine/threonine kinase, recent research has revealed it also functions as a scaffolding protein, involved in protein–protein interactions and organizing signaling complexes. One of the most well-studied examples of GSK-3 acting as a scaffolding protein is in the Wnt signaling pathway. GSK-3, Axin, adenomatous polyposis coli (APC), and casein kinase 1α (CK1α) forms a "destruction complex" without Wnt signaling, leading to the phosphorylation and subsequent degradation of β-catenin, a key effector of Wnt signaling. GSK-3 plays a crucial role in maintaining the stability of this complex, thereby enabling coordinated phosphorylation of β-catenin [14,16].

Furthermore, GSK-3 functions as a scaffold in the mTORC1 signaling pathway where it interacts with TSC2 (tuberous sclerosis complex 2) and PRAS40 (proline-rich Akt substrate of 40 kDa), two negative regulators of mTORC1. The interaction helps to inhibit mTORC1 activity under nutrient-poor conditions [17]. In addition to these findings, other studies have proved that GSK-3 can interact with various proteins in a kinase-independent manner, further demonstrating its role as a scaffolding protein [18].

The multifaceted roles of GSK-3 in the development of tumors and the progression of cancer make it a prime target for the development of new cancer treatments. Numerous GSK-3 inhibitors have demonstrated their potential in combating cancer through both preclinical and clinical studies [19,20]. One such example is Tideglusib, a non-ATP-competitive GSK-3β inhibitor. This inhibitor has been evaluated in xenograft and PDX murine models for a wide range of human cancers, including neuroblastoma [21], glioblastoma [22], osteosarcoma [23], prostate [24], pancreatic [25], and lung cancer [26].

TWS119 is another GSK-3β inhibitor and has demonstrated its potential in curtailing cell proliferation and triggering apoptosis in human alveolar rhabdomyosarcoma cells [27]. Moreover, it exhibits efficacy in effectively regulating the intricate processes of epithelial–mesenchymal transition (EMT) and cancer stem cell (CSC) properties in triple-negative breast cancer [28].

In the context of CAR-T cell therapy, GSK-3 inhibition has been found to enhance the persistence and antitumor activity of CAR-T cells. In a study in mantle cell lymphoma, the innovative CD19-CD22 bispecific CAR-T cells, when augmented with TWS119, displayed a remarkable reduction in the expression of exhaustion markers such as PD-1 and LAG3. The combination of bispecific CAR-T cells with TWS119 not only exhibits enhanced antitumor capabilities in vitro but also outperforms both the untreated control group and the groups treated with single CAR constructs in vivo studies. The combination of GSK-3 inhibitors with CAR-T cell therapy may provide a double therapeutic advantage by directly targeting the tumor cells and promoting a more effective immune response mediated by the engineered CAR-T cells [29].

2.2. GAPDH

GAPDH (glyceraldehyde-3-phosphate dehydrogenase) is a ubiquitous enzyme central to glycolysis and has garnered attention for its multifaceted role in cancer biology. GAPDH demonstrates both enzymatic and non-enzymatic functions, with both being intricately involved in cancer progression, invasiveness, and metastasis [30]. The enzymatic function of GAPDH is mainly observed in glycolysis, where it catalyzes the conversion of glyceraldehyde-3-phosphate to 1,3-bisphosphoglycerate. This function is important for cancer cells as it provides energy to support rapid proliferation and survival within the tumor microenvironment [31,32].

Furthermore, GAPDH's non-enzymatic functions are also critical in cancer progression by influencing diverse cellular processes beyond metabolism. Studies have shown that GAPDH is involved in transcriptional regulation [33,34], DNA repair [35,36], apoptosis modulation [37–39], and intracellular trafficking [40,41]. For example, apoptotic stimuli can activate NO formation and lead to the S-nitrosylation of GAPDH. This modification inhibits GAPDH's catalytic activity and enables it to bind to Siah, an E3-ubiquitin-ligase, which transports GAPDH to the nucleus. Inside the nucleus, GAPDH stabilizes Siah, allowing it to degrade specific target proteins and influence apoptosis. In the context of DNA repair, it is found that the activation of the tyrosine kinase Src occurs under DNA damage stress, leading to the phosphorylation of GAPDH at Tyr41. This phosphorylation event is crucial for the nuclear translocation of GAPDH. Nuclear GAPDH is recruited to DNA lesions and interacts with DNA polymerase β (Pol β) to participate in DNA repair mechanisms. Furthermore, nuclear GAPDH enhances Pol β polymerase activity and improves the efficiency of base excision repair (BER). GAPDH is also involved in apoptosis cascade in the aging process where the S-nitrosation of GAPDH, induced by iNOS, triggers its relocation to the nucleus. Once there, GAPDH actively mediates the process of apoptosis. These non-metabolic functions significantly promote tumorigenesis and tumor progression [31,32].

The dual roles of GAPDH in cancer are a good example of its ability to control cell death pathways. While some studies suggest GAPDH can increase cell apoptosis, others describe a protective function that promotes cell survival and tumor progression, which is extensively reviewed in [31]. Further exploration of the intricate interplay between GAPDH's metabolic and non-metabolic functions is helpful for us to better understand how it is involved in tumor initiation and progression. This is essential for designing innovative therapeutic strategies targeting GAPDH.

2.3. ENO1

Enolase, known as phosphopyruvate hydratase, is another example of a moonlighting enzyme. Enolase catalyzes the conversion of 2-phosphoglycerate (2-PG) into phosphoenolpyruvate (PEP), the ninth step of glycolysis. Enolase is essential in energy metabolism and is expressed in all tissues and organisms [42,43].

There are three enolase subunits in humans: α, β, and γ, encoded by separate genes. The three subunits can combine into five isoenzymes. Among these, alpha-enolase (ENO1) plays a significant role in cancer progression, acting not only as an enzymatic protein, but also as a plasminogen receptor on the cell surface [44–46]. As a plasminogen receptor on the cell surface, ENO1 facilitates the conversion of plasminogen to plasmin, a potent proteolytic enzyme that degrades the extracellular matrix (ECM) and, thus, facilitates cell migration [45]. Studies have also shown that ENO1 expression correlates with poor prognosis of cancer patients, implicating a potential as a prognostic biomarker [45]. Furthermore,

ENO1 participates in regulating integrin expression, which is a critical player in cancer cell adhesion, invasion, and metastasis [47]. The interaction between ENO1 and other proteins, such as protein arginine methyltransferase 5 (PRMT5), further exhibits its impact on cancer [48].

Overall, ENO1's dual functions as an enzyme and plasminogen receptor underscore its significance in cancer progression and highlight its potential as a therapeutic target for cancer treatment.

2.4. PTP4A3

Another typical moonlighting protein falling into this category is PTP4A3, also known as protein tyrosine phosphatase 4A3 or PRL-3. It is a member of the protein tyrosine phosphatase (PTP) family. As its name suggests, PTP4A3 catalyzes the dephosphorylation of tyrosine or serine/threonine residues in protein molecules, playing a crucial role in cellular signaling transduction. Its role in promoting cancer initiation, metastasis, drug resistance, and recurrence has been validated in various cancer cell lines and animal models, with its high expression being inversely correlated with the prognosis of several solid tumors [49]. Several substrates of PTP4A3 have been reported, including FZR1 [50], Keratin 8 [51,52], Integrin β1 [52], and Leo 1 [53], where the phosphorylation levels of the substrates are regulated by PTP4A3, thereby affecting cancer-related signaling pathways and cancer progression.

On the other hand, there is substantial experimental evidence indicating that PTP4A3 functions as a "pseudophosphatase". For instance, PTP4A3 can bind to CNNM4 (a membrane protein related to magnesium ion transport), inhibiting the efflux of Mg^{2+} and thereby impacting downstream energy metabolism pathways to promote tumorigenesis and progression [54]. The crystal structure of the PTP4A3-CNNM3 complex reveals that PTP4A3 acts as a "pseudophosphatase"; the critical amino acid residue C104 in the phosphatase active center of PTP4A3 binds to Asp426 of CNNM3, which is not phosphorylated [55,56].

In another study on breast cancer, it was found that PTP4A3 can induce the transformation of normal cancer cells into stem-like cells by competitively binding to MEF2A with HDAC4 (histone deacetylase 4), a process independent of PTP4A3's phosphatase activity. The interaction between PTP4A3 and HDAC4 leads to the dissociation of HDAC4 from MEF2A and histones, affecting the deacetylation reaction. Maintaining high levels of acetylated MEF2A and histones upregulates the expression of the stem cell key transcription factor SOX2, ultimately inducing cancer cell transformation into stem-like cells. Researchers have defined this function of PTP4A3 as an adaptor protein [57].

Our research in acute T-cell lymphoblastic leukemia (T-ALL) has shown that although PTP4A3 can influence the phosphorylation levels of Src kinase [58], subsequent studies have revealed that its oncogenic role in T-ALL does not depend on its enzymatic activity. The interactions between PTP4A3 and proteins such as LCK and CD3 in T-ALL are also independent of its enzymatic activity (unpublished data). Increasing evidence suggests that PTP4A3 is a moonlighting protein, concurrently performing the functions of a phosphatase and an adaptor.

Like GSK-3, the scaffolding role of PTP4A3 operates independently of its enzymatic action. However, the exact mechanism by which PTP4A3 toggles between its phosphatase activity and its "moonlighting" protein function within the context of diverse tumor types remains an area of active investigation. The dominant function that underpins its oncogenic role in tumors is yet to be conclusively determined.

3. Regulatory Mechanisms Governing Functions of Moonlighting Enzymes

Moonlighting enzymes are regulated by various mechanisms to ensure proper function switching. Post-translational modifications (PTMs) play a crucial role in prompting moonlighting proteins to switch functions. Additionally, the location and timing of each protein activity are vital. Moonlighting proteins may perform distinct functions in various cellular locations such as the cytoplasm, nucleus, or cell membrane.

3.1. *The Impact of Spatial Positioning on the Functions of Moonlighting Enzymes*

The location of proteins indeed plays a significant role in influencing their functions, and this principle applies to moonlighting enzymes as well. They may perform different roles in various cellular locations, such as the cytoplasm, nucleus, or cell membrane [59].

For example, the subcellular location of GAPDH significantly influences its functions, which is extensively reviewed in [60,61]. To encapsulate, GAPDH exhibits multifaceted roles depending on its cellular location. In the cytosol, it primarily performs as a pivotal enzyme in the metabolic process of glycolysis, catalyzing the conversion of glyceraldehyde 3-phosphate into 1,3-bisphosphoglycerate. However, when GAPDH translocates to the nucleus, it participates in the orchestration of gene expression, apoptosis, and DNA repair. At the cell surface, GAPDH can act as a receptor for certain bacterial pathogens, aiding in their entry into the cell. Within mitochondria, GAPDH contributes to apoptosis by interacting with voltage-dependent anion channels.

Another typical example of how subcellular location tightly regulates moonlighting functions of enzyme p53. This well-known tumor suppressor performs its canonical role as a transcription factor in the nucleus. Mono-ubiquitinated p53 induced by low levels of Mdm2 moves to the cytoplasm, where it triggers apoptosis and inhibits autophagy. The detailed regulation is reviewed in [62].

3.2. *Regulation via Post-Translational Modification (PTM)*

Post-translational modifications (PTMs) play an indispensable role in regulating protein functions and their subcellular localization. This universal principle is true for the regulation of moonlighting proteins as well. The ensuing discussion will focus on several key PTMs that significantly contribute to the nuanced regulation of moonlighting protein functions.

3.2.1. Phosphorylation

Phosphorylation is one of the most common PTMs. Phosphorylation of moonlighting proteins can modulate their activity levels or even switch between divergent functions. For example, serine residue phosphorylation of GAPDH inhibits its glycolytic function while enhancing nuclear translocation, where it regulates gene transcription [61].

In the case of the α-enolase isoform, phosphorylation of specific tyrosine residues in the protein promotes its interaction with plasminogen to cleave the activation peptide of plasminogen, leading to the conversion of plasminogen to plasmin, an enzymatic mediator implicated in extracellular matrix degradation and tissue remodeling [44]. These instances underscore phosphorylation's dynamic regulatory potential over moonlighting proteins.

3.2.2. Acetylation

Acetylation is a PTM modification adding an acetyl group to a protein, typically on lysine residues. Acetylation generally affects protein stability, protein–protein interactions, and subcellular localization, all of which are closely related to their activity. For instance, acetylation of Hsp90 at lysine residue 294 has been reported to enhance its interaction with transcription factors, such as p53, leading to altered gene expression patterns. This acetylation modification converts Hsp90 from a canonical chaperone to a transcriptional regulator [63].

3.2.3. Ubiquitination

Ubiquitination is a PTM where a small protein, ubiquitin, is attached to target proteins by a cascade of E3 enzymatic reactions. Ubiquitination can occur at different residues on the target protein, including lysine (K) residues, as well as the amino terminus (N-terminus) of the protein. This modification can regulate protein function, localization, or protein–protein interactions. Ubiquitination of moonlighting proteins can regulate the balance between their distinct functions by targeting them for degradation or influencing their interactions with other proteins, for example, a well-known tumor suppressor, PTEN, whose primary role is a lipid phosphatase. However, PTEN also has a less-understood nuclear function, where it engages in maintaining chromosomal integrity [64]. The ubiquitination of PTEN plays a crucial role in regulating its cellular localization, and thus its function. Mono-ubiquitinated PTEN translocates to the nucleus where it plays a role in DNA repair and maintaining chromosomal stability. Meanwhile, non-ubiquitinated PTEN resides in the cytoplasm and acts as a phosphatase to negatively regulate the PI3K/Akt pathway [65].

3.2.4. SUMOylation

Small ubiquitin-like modifier (SUMO) modification involves the attachment of SUMO proteins to target proteins. SUMOylation of moonlighting proteins can impact their functions by modulating their subcellular localization or interactions with other proteins. SUMOylation of p53 can switch its function from a transcription factor to a regulator of cellular metabolism. SUMOylation of p53 at specific lysine residues, such as K386 and K390, inhibits p53-dependent transcription by preventing its binding to DNA/chromatin [66]. When p53 is SUMOylated, it can translocate to the mitochondria and interact with proteins involved in metabolic pathways [66].

The intricate processes governing the regulation of moonlighting proteins' diverse functionalities largely remain enigmatic; for instance, the subcellular location of PTP4A3 and its PTM are largely unknown. PTP4A3 belongs to a class of prenylated protein tyrosine phosphatases (PTPs) that are associated with the cell plasma membrane. Studies have shown that these PTPs are prenylated proteins in vivo, suggesting their localization at the cell membrane [67]. Therefore, research focusing on the subcellular location and post-translational modifications (PTMs) of moonlighting enzymes, and how they regulate multiple-function switching, are indeed valuable areas of exploration.

4. Targeting Moonlighting Enzymes for Cancer Treatment

There are two main strategies to effectively target multifunctional enzymes. The first is precise targeting, which blocks a specific function while leaving the others intact. Consider an enzyme that is important to both standard metabolic functions and anti-apoptotic pathways. If the purpose is to block the enzyme's anti-apoptotic function without meddling with its metabolic role, then precise targeting would be the preferred choice.

The secondary strategy, termed as the "abolish-all" tactic, is designed to suppress all the enzyme's functions. This method is particularly beneficial where every activity is hijacked to advance the disease. By simply turning off everything, we can nullify the enzyme's multifunctionality and prevent all its harmful impacts in the disease progression.

In the realm of current drug design technologies, the strategy of precise targeting poses immense challenges. Therefore, our upcoming discussion will focus on the abolish-all strategy, which can be further subdivided into inhibiting the enzyme's protein expression, inhibiting enzyme activity, or target protein degradation (TPD). TPD is a strategy that has emerged in the past two decades, representing a more thorough abolish-all approach. Hence, we will now delve into the potential of this strategy in targeting dual-function enzymes, using PTP4A3 as a case study.

4.1. Target Protein Degradation

The traditional method of drug discovery primarily involves directly modulating protein activity. The development and application of protein activity regulators, particularly inhibitors, have long been the mainstream of drug development. Over the past two decades, the technology of proteolysis targeting chimeras (PROTACs), which utilizes the body's own protein clearance system to remove pathogenic target proteins, has been rapidly developing and has become a major weapon in new drug research and development. In addition to PROTACs, several new targeted protein degradation (TPD) strategies have emerged, including molecular glues, lysosome-targeting chimeras (LYTACs), and antibody-based PROTACs (AbTACs). TPD primarily utilizes the two major protein clearance systems in cells: the ubiquitin–proteasome system (UPS) and the lysosome. Most TPD strategies, such as PROTACs, molecular glues, and degradation tags (dTAGs), rely on the UPS and are mainly targeted at intracellular proteins. Lysosome-dependent TPD strategies can degrade membrane proteins, extracellular proteins, and protein aggregates, thereby greatly expanding the range of substrates. The history and current status of TPD development, as well as its application in drug discovery and design, have been summarized in numerous review articles [68–75].

The key difference between TPD, such as PROTAC technology, and traditional small molecule protein inhibitors/antibodies, is that PROTAC is "event driven", as opposed to small molecules or antibodies which are "occupancy driven". Occupancy driven implies that when small molecule inhibitors or antibodies inhibit a target protein's activity, they need to occupy the protein's active center for a long time. Small molecule inhibitors need to fulfill multiple requirements, including the following: a. high enough affinity to outcompete the target protein's natural ligand/receptor; b. a large enough dose to saturate the target; and c. a long enough half-life to persistently inhibit protein activity. Therefore, the target protein needs a good drug pocket to allow the drug to bind firmly and regulate its activity [76].

In contrast, PROTAC only needs to briefly bind with the target protein/E3 enzyme, thereby triggering the degradation event and qualifying as "event driven". It does not need to directly inhibit the function of the target protein, nor does it require sustained and intensive binding with the target protein. In theory, as long as the target protein has a ligand that can bind briefly, it can target proteins without suitable drug pockets or without an enzymatic active center. It has also been found in practice that PROTAC has a significant advantage in being able to transform specific poor target protein ligands into specific strong bifunctional molecules. For example, Foretinib is a pan-kinase inhibitor that can act on more than one hundred kinases, while two PROTAC molecules based on Foretinib can only degrade 14 and 9 kinases, respectively, greatly increasing their specificity. The reason for this might be that the interaction between target protein and E3 ubiquitin enzyme enhances selectivity [77,78].

4.2. Degrading PTP4A3 for Cancer Therapy

As discussed in Section 2.4, PTP4A3 is a recently discovered cancer target. Besides possessing traditional phosphatase functions, it also acts as a scaffolding protein or pseudophosphatase, and these functions are related to its cancer-promoting effects. The development of small molecule drugs targeting PTP4A3 faces multiple challenges. The enzyme active center, i.e., the binding cavity for small molecule compounds, is very flat and narrow, making it difficult to screen for high-affinity small molecule compounds [49]. Additionally, it is generally more difficult to screen for compounds that interfere with protein non-enzyme-promoting functions than to screen for compounds that inhibit its enzymatic activity.

Much effort has been put into developing PTP4A3 inhibitors. Various lead compound screening strategies, including high-throughput screening, virtual screening, and natural product screening, have been used in the research and development of PTP4A3 inhibitors. More than a dozen small molecule compounds that can inhibit the phosphatase activity of PTP4A3 in vitro have been discovered. These include JMS-053 [79], thienopyridone [80], analog 3 [81,82], rhodanine and its derivatives [83–85], and various natural products [49]. JMS-053, thienopyridone, analog 3, and rhodanine are widely used as tool drugs in the basic research of PTP4A3. However, most of these small molecule inhibitors have limited therapeutic effects or poor selectivity [49]. Therefore, no PTP4A3 small molecule inhibitors have entered clinical research yet.

Thus, a new horizon in overcoming the challenges of poor drug-like properties could be the development of PROTAC molecules that specifically target PTP4A3. This could pave the way for a new era of clinical drug development, specifically for the treatment of malignant tumors that are PTP4A3-dependent. We recently carried out a principle of concept study of developing a nanobody-based bio-PROTAC for the purpose of degrading PTP4A3.

The preliminary study showed that this bio-PROTAC could degrade GFP-PTP4A3 fusion protein while sparing the homologous GFP-PTP4A1 or GFP-PTP4A2 (unpublished data). These preliminary research findings illuminate a path towards the targeted degradation of PTP4A3. We aim to evaluate the role of this bio-PROTAC in cancer.

5. Future Perspectives

It is now widely accepted that moonlighting is a universe nature of proteins, which is also true for enzymes. However, when moonlighting enzymes come under our scrutiny, particularly those with both enzymatic and non-enzymatic activities, we often neglect their non-enzymatic roles. Perhaps the exhaustive examination of their catalytic functions presented in the literature has generated a stereotype. However, such oversight could potentially pose limitations for drug design and development. For enzymes of this nature, it might be beneficial to consider an "abolish-all" approach, a strategy discussed above.

Author Contributions: Conceptualization, M.W. and L.Z.; software, C.L.; investigation, C.L.; resources, X.W.; writing—original draft preparation, C.L. and M.Y.; writing—review and editing, M.W.; visualization, H.W.; supervision, M.W.; project administration, L.Z.; funding acquisition, M.W. All authors have read and agreed to the published version of the manuscript.

Funding: The research was funded by Science Foundation of Guangzhou Science and Technology Bureau: 202201010035.

Data Availability Statement: No new data were created.

Conflicts of Interest: The authors declare no conflict of interest.

References

1. Jeffery, C.J. Moonlighting proteins. *Trends Biochem. Sci.* **1999**, *24*, 8–11. [CrossRef]
2. Singh, N.; Bhalla, N. Moonlighting Proteins. *Annu. Rev. Genet.* **2020**, *54*, 265–285. [CrossRef]
3. Jeffery, C.J. Protein species and moonlighting proteins: Very small changes in a protein's covalent structure can change its biochemical function. *J. Proteom.* **2016**, *134*. [CrossRef]
4. Jeffery, C.J. Enzymes, pseudoenzymes, and moonlighting proteins: Diversity of function in protein superfamilies. *FEBS J.* **2020**, *287*, 4141–4149. [CrossRef] [PubMed]
5. Huberts, D.H.E.W.; van der Klei, I.J. Moonlighting proteins: An intriguing mode of multitasking. *Biochim. Biophys. Acta-Mol. Cell Res.* **2010**, *1803*. [CrossRef] [PubMed]
6. Min, K.W.; Lee, S.H.; Baek, S.J. Moonlighting proteins in cancer. *Cancer Lett.* **2016**, *370*, 108–116. [CrossRef] [PubMed]
7. Jeffery, C.J. Protein moonlighting: What is it, and why is it important? *Philos. Trans. R. Soc. B Biol. Sci.* **2018**, *373*, 20160523. [CrossRef] [PubMed]
8. Espinosa-Cantú, A.; Ascencio, D.; Barona-Gómez, F.; De Luna, A. Gene duplication and the evolution of moonlighting proteins. *Front. Genet.* **2015**, *6*, 227. [CrossRef]
9. Adamo, A.; Frusteri, C.; Pallotta, M.T.; Pirali, T.; Sartoris, S.; Ugel, S. Moonlighting Proteins Are Important Players in Cancer Immunology. *Front. Immunol.* **2021**, *11*, 613069. [CrossRef]
10. Jeffery, C.J. An enzyme in the test tube, and a transcription factor in the cell: Moonlighting proteins and cellular factors that affect their behavior. *Protein Sci.* **2019**, *28*, 1233–1238. [CrossRef]
11. Bakheet, T.M.; Doig, A.J. Properties and identification of human protein drug targets. *Bioinformatics* **2009**, *25*, 451–457. [CrossRef] [PubMed]
12. Kazi, A.; Xiang, S.; Yang, H.; Delitto, D.; Trevino, J.; Jiang, R.H.Y.; Ayaz, M.; Lawrence, H.R.; Kennedy, P.; Sebti, S.M. GSK3 suppression upregulates β-catenin and c-Myc to abrogate KRas-dependent tumors. *Nat. Commun.* **2018**, *9*, 5154. [CrossRef] [PubMed]
13. Fang, X.; Yu, S.X.; Lu, Y.; Bast, R.C.; Woodgett, J.R.; Mills, G.B. Phosphorylation and inactivation of glycogen synthase kinase 3 by protein kinase A. *Proc. Natl. Acad. Sci. USA* **2000**, *97*, 11960–11965. [CrossRef] [PubMed]
14. Stamos, J.L.; Weis, W.I. The β-catenin destruction complex. *Cold Spring Harb. Perspect. Biol.* **2013**, *5*, a007898. [CrossRef] [PubMed]
15. Sears, R.; Nuckolls, F.; Haura, E.; Taya, Y.; Tamai, K.; Nevins, J.R. Multiple Ras-dependent phosphorylation pathways regulate Myc protein stability. *Genes. Dev.* **2000**, *14*, 2501–2514. [CrossRef]
16. Kimelman, D.; Xu, W. β-Catenin destruction complex: Insights and questions from a structural perspective. *Oncogene* **2006**, *25*. [CrossRef] [PubMed]
17. Inoki, K.; Li, Y.; Zhu, T.; Wu, J.; Guan, K.L. TSC2 is phosphorylated and inhibited by Akt and suppresses mTOR signalling. *Nat. Cell Biol.* **2002**, *4*, 648–657. [CrossRef]

18. Beurel, E.; Grieco, S.F.; Jope, R.S. Glycogen synthase kinase-3 (GSK3): Regulation, actions, and diseases. *Pharmacol. Ther.* **2015**, *148*, 114–131. [CrossRef]
19. Mathuram, T.L.; Reece, L.M.; Cherian, K.M. GSK-3 Inhibitors: A Double-Edged Sword? An Update on Tideglusib. *Drug Res.* **2018**, *68*, 436–443. [CrossRef]
20. Augello, G.; Emma, M.R.; Cusimano, A.; Azzolina, A.; Montalto, G.; McCubrey, J.A.; Cervello, M. The role of GSK-3 in cancer immunotherapy: GSK-3 inhibitors as a new frontier in cancer treatment. *Cells* **2020**, *9*, 1427. [CrossRef]
21. Bahmad, H.F.; Chalhoub, R.M.; Harati, H.; Bou-Gharios, J.; Assi, S.; Ballout, F.; Monzer, A.; Msheik, H.; Araji, T.; Elajami, M.K.; et al. Tideglusib attenuates growth of neuroblastoma cancer stem/progenitor cells in vitro and in vivo by specifically targeting GSK-3β. *Pharmacol. Rep.* **2021**, *73*, 211–226. [CrossRef] [PubMed]
22. Bou-Gharios, J.; Assi, S.; Bahmad, H.F.; Kharroubi, H.; Araji, T.; Chalhoub, R.M.; Ballout, F.; Harati, H.; Fares, Y.; Abou-Kheir, W. The potential use of tideglusib as an adjuvant radio-therapeutic treatment for glioblastoma multiforme cancer stem-like cells. *Pharmacol. Rep.* **2021**, *73*, 227–239. [CrossRef] [PubMed]
23. Wei, D.; Zhu, X.; Li, S.; Liu, G.; Wang, Y.; Wang, W.; Zhang, Q.; Jiang, S. Tideglusib suppresses stem-cell-like features and progression of osteosarcoma by inhibiting GSK-3β/NOTCH1 signaling. *Biochem. Biophys. Res. Commun.* **2021**, *554*, 206–213. [CrossRef]
24. Sun, A.; Li, C.; Chen, R.; Huang, Y.; Chen, Q.; Cui, X.; Liu, H.; Thrasher, J.B.; Li, B. GSK-3β controls autophagy by modulating LKB1-AMPK pathway in prostate cancer cells. *Prostate* **2016**, *76*, 172–183. [CrossRef] [PubMed]
25. Hao, Q.; Gao, L.; Niu, W.; Chen, L.; Zhang, P.; Chen, Z. POTEE stimulates the proliferation of pancreatic cancer by activating the PI3K/Akt/GSK-3β/β-catenin signaling. *BioFactors* **2020**, *46*, 685–692. [CrossRef] [PubMed]
26. Daouk, R.; Hassane, M.; Bahmad, H.F.; Sinjab, A.; Fujimoto, J.; Abou-Kheir, W.; Kadara, H. Genome-wide and phenotypic evaluation of stem cell progenitors derived from GPRC5A-deficient murine lung adenocarcinoma with somatic KRAS mutations. *Front. Oncol.* **2019**, *9*, 207. [CrossRef]
27. Zeng, F.Y.; Dong, H.; Cui, J.; Liu, L.; Chen, T. Glycogen synthase kinase 3 regulates PAX3-FKHR-mediated cell proliferation in human alveolar rhabdomyosarcoma cells. *Biochem. Biophys. Res. Commun.* **2010**, *391*, 1049–1055. [CrossRef]
28. Vijay, G.V.; Zhao, N.; Den Hollander, P.; Toneff, M.J.; Joseph, R.; Pietila, M.; Taube, J.H.; Sarkar, T.R.; Ramirez-Pena, E.; Werden, S.J.; et al. GSK3β regulates epithelial-mesenchymal transition and cancer stem cell properties in triple-negative breast cancer. *Breast Cancer Res.* **2019**, *21*, 37. [CrossRef]
29. Jin, J.; Li, Y.; Liu, Y.; Jordan, A.A.; McIntosh, J.; Vargas, J.; Che, Y.; Yao, Y.; Wang, M. Bispecific CD19-CD20 and CD19-CD22 CAR-T Cells with Glycogen Synthase Kinase (GSK)-3β Inhibitor TWS119 Treatment Have Superior Therapeutic Effects on Mantle Cell Lymphoma. *Blood* **2021**, *138* (Suppl. 1), 1698. [CrossRef]
30. Sirover, M.A. Pleiotropic effects of moonlighting glyceraldehyde-3-phosphate dehydrogenase (GAPDH) in cancer progression, invasiveness, and metastases. In *Cancer and Metastasis Reviews*; Springer: New York, NY, USA, 2018; Volume 37, pp. 665–676. [CrossRef]
31. Colell, A.; Green, D.R.; Ricci, J.E. Novel roles for GAPDH in cell death and carcinogenesis. *Cell Death Differ.* **2009**, *16*, 1573–1581. [CrossRef]
32. Zhang, J.Y.; Zhang, F.; Hong, C.Q.; Giuliano, A.E.; Cui, X.J.; Zhou, G.J.; Zhang, G.J.; Cui, Y.K. Critical protein GAPDH and its regulatory mechanisms in cancer cells. *Cancer Biol. Med.* **2015**, *12*, 10–22. [CrossRef] [PubMed]
33. Hara, M.R.; Snyder, S.H. Nitric oxide-GAPDH-Siah: A novel cell death cascade. *Cell. Mol. Neurobiol.* **2006**, *26*, 525–536. [CrossRef]
34. Sun, X.H.; Lis, J.T.; Wu, R. The positive and negative transcriptional regulation of the Drosophila Gapdh-2 gene. *Genes. Dev.* **1988**, *2*, 743–753. [CrossRef] [PubMed]
35. Kosova, A.A.; Khodyreva, S.N.; Lavrik, O.I. Role of glyceraldehyde-3-phosphate dehydrogenase (GAPDH) in DNA repair. *Biochemistry* **2017**, *82*, 643–654. [CrossRef] [PubMed]
36. Ci, S.; Xia, W.; Liang, W.; Qin, L.; Zhang, Y.; Dianov, G.L.; Wang, M.; Zhao, X.; Wu, C.; Alagamuthu, K.K.; et al. Src-mediated phosphorylation of GAPDH regulates its nuclear localization and cellular response to DNA damage. *FASEB J.* **2020**, *34*, 10443–10461. [CrossRef] [PubMed]
37. Chen, R.W.; Saunders, P.A.; Wei, H.; Li, Z.; Seth, P.; Chuang, D.M. Involvement of glyceraldehyde-3-phosphate dehydrogenase (GAPDH) and p53 in neuronal apoptosis: Evidence that GAPDH is upregulated by p53. *J. Neurosci.* **1999**, *19*, 9654–9662. [CrossRef]
38. Xie, T.; Qiao, X.; Sun, C.; Chu, B.; Meng, J.; Chen, C. GAPDH S-nitrosation contributes to age-related sarcopenia through mediating apoptosis. *Nitric Oxide* **2022**, *120*, 1–8. [CrossRef]
39. Hou, X.; Snarski, P.; Higashi, Y.; Yoshida, T.; Jurkevich, A.; Delafontaine, P.; Sukhanov, S. Nuclear complex of glyceraldehyde-3-phosphate dehydrogenase and DNA repair enzyme apurinic/apyrimidinic endonuclease i protect smooth muscle cells against oxidant-induced cell death. *FASEB J.* **2017**, *31*, 3179–3192. [CrossRef]
40. Tisdale, E.J.; Talati, N.K.; Artalejo, C.R.; Shisheva, A. GAPDH binds Akt to facilitate cargo transport in the early secretory pathway. *Exp. Cell Res.* **2016**, *349*, 310–319. [CrossRef]
41. Tisdale, E.J.; Kelly, C.; Artalejo, C.E. Glyceraldehyde-3-phosphate dehydrogenase interacts with Rab2 and plays an essential role in endoplasmic reticulum to golgi transport exclusive of its glycolytic activity. *J. Biol. Chem.* **2004**, *279*, 54046–54052. [CrossRef]
42. Díaz-Ramos, À.; Roig-Borrellas, A.; García-Melero, A.; López-Alemany, R. α-enolase, a multifunctional protein: Its role on pathophysiological situations. *J. Biomed. Biotechnol.* **2012**, *2012*, 156795. [CrossRef] [PubMed]

43. Didiasova, M.; Schaefer, L.; Wygrecka, M. When place matters: Shuttling of enolase-1 across cellular compartments. *Front. Cell Dev. Biol.* **2019**, *7*, 61. [CrossRef] [PubMed]
44. Pancholi, V. Multifunctional α-enolase: Its role in diseases. *Cell. Mol. Life Sci.* **2001**, *58*, 902–920. [CrossRef] [PubMed]
45. Almaguel, F.A.; Sanchez, T.W.; Ortiz-Hernandez, G.L.; Casiano, C.A. Alpha-Enolase: Emerging Tumor-Associated Antigen, Cancer Biomarker, and Oncotherapeutic Target. *Front. Genet.* **2021**, *11*, 614726. [CrossRef] [PubMed]
46. Song, Q.; Zhang, K.; Sun, T.; Xu, C.; Zhao, W.; Zhang, Z. Knockout of ENO1 leads to metabolism reprogramming and tumor retardation in pancreatic cancer. *Front. Oncol.* **2023**, *13*, 1119886. [CrossRef] [PubMed]
47. Principe, M.; Borgoni, S.; Cascione, M.; Chattaragada, M.S.; Ferri-Borgogno, S.; Capello, M.; Bulfamante, S.; Chapelle, J.; Di Modugno, F.; Defilippi, P.; et al. Alpha-enolase (ENO1) controls alpha v/beta 3 integrin expression and regulates pancreatic cancer adhesion, invasion, and metastasis. *J. Hematol. Oncol.* **2017**, *10*, 16. [CrossRef] [PubMed]
48. Zakrzewicz, D.; Didiasova, M.; Krüger, M.; Giaimo, B.D.; Borggrefe, T.; Mieth, M.; Hocke, A.C.; Zakrzewicz, A.; Schaefer, L.; Preissner, K.T.; et al. Protein arginine methyltransferase 5 mediates enolase-1 cell surface trafficking in human lung adenocarcinoma cells. *Biochim. Biophys. Acta Mol. Basis Dis.* **2018**, *1864*, 1816–1827. [CrossRef] [PubMed]
49. Wei, M.; Korotkov, K.V.; Blackburn, J.S. Targeting phosphatases of regenerating liver (PRLs) in cancer. *Pharmacol. Ther.* **2018**, *190*, 128–138. [CrossRef] [PubMed]
50. Zhang, C.; Qu, L.; Lian, S.; Meng, L.; Min, L.; Liu, J.; Song, Q.; Shen, L.; Shou, C. PRL-3 promotes ubiquitination and degradation of AURKA and colorectal cancer progression via dephosphorylation of FZR1. *Cancer Res.* **2019**, *79*, 928–940. [CrossRef]
51. Mizuuchi, E.; Semba, S.; Kodama, Y.; Yokozaki, H. Down-modulation of keratin 8 phosphorylation levels by PRL-3 contributes to colorectal carcinoma progression. *Int. J. Cancer* **2009**, *124*, 1802–1810. [CrossRef]
52. Peng, L.; Xing, X.; Li, W.; Qu, L.; Meng, L.; Lian, S.; Jiang, B.; Wu, J.; Shou, C. PRL-3 promotes the motility, invasion, and metastasis of LoVo colon cancer cells through PRL-3-integrin β1-ERK1/2 and-MMP2 signaling. *Mol. Cancer* **2009**, *8*, 110. [CrossRef] [PubMed]
53. Chong, P.S.Y.; Zhou, J.; Cheong, L.L.; Liu, S.C.; Qian, J.; Guo, T.; Sze, S.K.; Zeng, Q.; Chng, W.J. LEO1 is regulated by PRL-3 and mediates its oncogenic properties in acute myelogenous leukemia. *Cancer Res.* **2014**, *74*, 3043–3053. [CrossRef] [PubMed]
54. Funato, Y.; Yamazaki, D.; Mizukami, S.; Du, L.; Kikuchi, K.; Miki, H. Membrane protein CNNM4-dependent Mg2+ efflux suppresses tumor progression. *J. Clin. Investig.* **2014**, *124*, 5398–5410. [CrossRef] [PubMed]
55. Zhang, H.; Kozlov, G.; Li, X.; Wu, H.; Gulerez, I.; Gehring, K. PRL3 phosphatase active site is required for binding the putative magnesium transporter CNNM3. *Sci. Rep.* **2017**, *7*, 48. [CrossRef] [PubMed]
56. Smith, C.N.; Kihn, K.; Williamson, Z.A.; Chow, K.M.; Hersh, L.B.; Korotkov, K.V.; Deredge, D.; Blackburn, J.S. Development and characterization of nanobodies that specifically target the oncogenic Phosphatase of Regenerating Liver-3 (PRL-3) and impact its interaction with a known binding partner, CNNM3. *PLoS ONE* **2023**, *18*, e0285964. [CrossRef] [PubMed]
57. Zhang, M.; Wei, Y.; Liu, Y.; Guan, W.; Zhang, X.; Kong, J.; Li, H.; Yang, S.; Wang, H. Metastatic Phosphatase PRL-3 Induces Ovarian Cancer Stem Cell Sub-population through Phosphatase-Independent Deacetylation Modulations. *iScience* **2020**, *23*, 100766. [CrossRef]
58. Wei, M.; Haney, M.G.; Rivas, D.R.; Blackburn, J.S. Protein tyrosine phosphatase 4A3 (PTP4A3/PRL-3) drives migration and progression of T-cell acute lymphoblastic leukemia in vitro and in vivo. *Oncogenesis* **2020**, *9*, 6. [CrossRef] [PubMed]
59. Gupta, M.N.; Uversky, V.N. Moonlighting enzymes: When cellular context defines specificity. *Cell. Mol. Life Sci.* **2023**, *80*, 130. [CrossRef] [PubMed]
60. Hildebrandt, T.; Knuesting, J.; Berndt, C.; Morgan, B.; Scheibe, R. Cytosolic thiol switches regulating basic cellular functions: GAPDH as an information hub? *Biol. Chem.* **2015**, *396*, 523–537. [CrossRef]
61. Tristan, C.; Shahani, N.; Sedlak, T.W.; Sawa, A. The diverse functions of GAPDH: Views from different subcellular compartments. *Cell. Signal.* **2011**, *23*, 317–323. [CrossRef]
62. Lee, J.T.; Gu, W. The multiple levels of regulation by p53 ubiquitination. *Cell Death Differ.* **2010**, *17*, 86–92. [CrossRef] [PubMed]
63. Scroggins, B.T.; Robzyk, K.; Wang, D.; Marcu, M.G.; Tsutsumi, S.; Beebe, K.; Cotter, R.J.; Felts, S.; Toft, D.; Karnitz, L.; et al. An Acetylation Site in the Middle Domain of Hsp90 Regulates Chaperone Function. *Mol. Cell* **2007**, *25*, 151–159. [CrossRef]
64. Shen, W.H.; Balajee, A.S.; Wang, J.; Wu, H.; Eng, C.; Pandolfi, P.P.; Yin, Y. Essential Role for Nuclear PTEN in Maintaining Chromosomal Integrity. *Cell* **2007**, *128*, 157–170. [CrossRef] [PubMed]
65. Trotman, L.C.; Wang, X.; Alimonti, A.; Chen, Z.; Teruya-Feldstein, J.; Yang, H.; Pavletich, N.P.; Carver, B.S.; Cordon-Cardo, C.; Erdjument-Bromage, H.; et al. Ubiquitination Regulates PTEN Nuclear Import and Tumor Suppression. *Cell* **2007**, *128*, 141–156. [CrossRef] [PubMed]
66. Wu, S.Y.; Chiang, C.M. Crosstalk between sumoylation and acetylation regulates p53-dependent chromatin transcription and DNA binding. *EMBO J.* **2009**, *28*, 1246–1259. [CrossRef] [PubMed]
67. Wu, S.Y.; Chiang, C.M. p53 sumoylation: Mechanistic insights from reconstitution studies. *Epigenetics* **2009**, *4*, 445–451. [CrossRef] [PubMed]
68. Hu, Z.; Crews, C.M. Recent Developments in PROTAC-Mediated Protein Degradation: From Bench to Clinic. *ChemBioChem* **2022**, *23*, e202100270. [CrossRef] [PubMed]
69. Alabi, S.B.; Crews, C.M. Major advances in targeted protein degradation: PROTACs, LYTACs, and MADTACs. *J. Biol. Chem.* **2021**, *296*, 100647. [CrossRef] [PubMed]
70. Li, K.; Crews, C.M. PROTACs: Past, present and future. *Chem. Soc. Rev.* **2022**, *51*, 5214–5236. [CrossRef]

71. Poso, A. The Future of Medicinal Chemistry, PROTAC, and Undruggable Drug Targets. *J. Med. Chem.* **2021**, *64*, 10680–10681. [CrossRef]
72. Sakamoto, K.M.; Kim, K.B.; Kumagai, A.; Mercurio, F.; Crews, C.M.; Deshaies, R.J. Protacs: Chimeric molecules that target proteins to the Skp1-Cullin-F box complex for ubiquitination and degradation. *Proc. Natl. Acad. Sci. USA* **2001**, *98*. [CrossRef] [PubMed]
73. Neklesa, T.K.; Winkler, J.D.; Crews, C.M. Targeted protein degradation by PROTACs. *Pharmacol. Ther.* **2017**, *174*, 138–144. [CrossRef]
74. Békés, M.; Langley, D.R.; Crews, C.M. PROTAC targeted protein degraders: The past is prologue. *Nat. Rev. Drug Discov.* **2022**, *21*, 181–200. [CrossRef] [PubMed]
75. Song, J.; Hu, M.; Zhou, J.; Xie, S.; Li, T.; Li, Y. Targeted protein degradation in drug development: Recent advances and future challenges. *Eur. J. Med. Chem.* **2023**, *261*, 115839. [CrossRef] [PubMed]
76. Pettersson, M.; Crews, C.M. PROteolysis TArgeting Chimeras (PROTACs)—Past, present and future. *Drug Discov. Today Technol.* **2019**, *31*, 15–27. [CrossRef] [PubMed]
77. Cromm, P.M.; Samarasinghe, K.T.G.; Hines, J.; Crews, C.M. Addressing Kinase-Independent Functions of Fak via PROTAC-Mediated Degradation. *J. Am. Chem. Soc.* **2018**, *140*, 17019–17026. [CrossRef] [PubMed]
78. Bondeson, D.P.; Smith, B.E.; Burslem, G.M.; Buhimschi, A.D.; Hines, J.; Jaime-Figueroa, S.; Wang, J.; Hamman, B.D.; Ishchenko, A.; Crews, C.M. Lessons in PROTAC Design from Selective Degradation with a Promiscuous Warhead. *Cell Chem. Biol.* **2018**, *25*, 78–87.e5. [CrossRef]
79. Lazo, J.S.; Isbell, K.N.; Vasa, S.A.; Llaneza, D.C.; Rastelli, E.J.; Wipf, P.; Sharlow, E.R. Disruption of Ovarian Cancer STAT3 and p38 Signaling with a Small-Molecule Inhibitor of PTP4A3 Phosphatase. *J. Pharmacol. Exp. Ther.* **2023**, *384*, 429–438.e5. [CrossRef] [PubMed]
80. Zhang, Z.; Kozlov, G.; Chen, Y.S.; Gehring, K. Mechanism of thienopyridone and iminothienopyridinedione inhibition of protein phosphatases. *Medchemcomm* **2019**, *10*, 791–799. [CrossRef]
81. Hoeger, B.; Diether, M.; Ballester, P.J.; Köhn, M. Biochemical evaluation of virtual screening methods reveals a cell-active inhibitor of the cancer-promoting phosphatases of regenerating liver. *Eur. J. Med. Chem.* **2014**, *88*, 89–100. [CrossRef]
82. Park, H.; Jung, S.K.; Jeong, D.G.; Ryu, S.E.; Kim, S.J. Discovery of novel PRL-3 inhibitors based on the structure-based virtual screening. *Bioorg. Med. Chem. Lett.* **2008**, *18*, 2250–2255. [CrossRef] [PubMed]
83. Min, G.; Lee, S.K.; Kim, H.N.; Han, Y.M.; Lee, R.H.; Jeong, D.G.; Han, D.C.; Kwon, B.M. Rhodanine-based PRL-3 inhibitors blocked the migration and invasion of metastatic cancer cells. *Bioorg. Med. Chem. Lett.* **2013**, *23*, 3769–3774. [CrossRef] [PubMed]
84. Ahn, J.H.; Kim, S.J.; Park, W.S.; Cho, S.Y.; Ha, J.D.; Kim, S.S.; Kang, S.K.; Jeong, D.G.; Jung, S.K.; Lee, S.H.; et al. Synthesis and biological evaluation of rhodanine derivatives as PRL-3 inhibitors. *Bioorg. Med. Chem. Lett.* **2006**, *16*, 2996–2999. [CrossRef]
85. Lin, L.; Lu, L.; Yuan, C.; Wang, A.; Zhu, M.; Fu, X.; Xing, S. The dual inhibition against the activity and expression of tyrosine phosphatase PRL-3 from a rhodanine derivative. *Bioorg. Med. Chem. Lett.* **2021**, *41*, 127981. [CrossRef] [PubMed]

MDPI AG
Grosspeteranlage 5
4052 Basel
Switzerland
Tel.: +41 61 683 77 34

Molecules Editorial Office
E-mail: molecules@mdpi.com
www.mdpi.com/journal/molecules

www.ingramcontent.com/pod-product-compliance
Lightning Source LLC
Chambersburg PA
CBHW062044220926
51850CB00010B/2659